D1325445

Undergraduate Lecture Notes in Physics

For further volumes:
http://www.springer.com/series/8917

Undergraduate Lecture Notes in Physics (ULNP) publishes authoritative texts covering topics throughout pure and applied physics. Each title in the series is suitable as a basis for undergraduate instruction, typically containing practice problems, worked examples, chapter summaries, and suggestions for further reading.

ULNP titles must provide at least one of the following:

- An exceptionally clear and concise treatment of a standard undergraduate subject.
- A solid undergraduate-level introduction to a graduate, advanced, or non-standard subject.
- A novel perspective or an unusual approach to teaching a subject.

ULNP especially encourages new, original, and idiosyncratic approaches to physics teaching at the undergraduate level.

The purpose of ULNP is to provide intriguing, absorbing books that will continue to be the reader's preferred reference throughout their academic career.

José Natário

General Relativity Without Calculus

A Concise Introduction to the Geometry of Relativity

 Springer

José Natário
Departamento de Matematica
Instituto Superior Tecnico
Av. Rovisco Pais
1049-001 Lisbon
Portugal
e-mail: jnatar@math.ist.utl.pt

ISSN 2192-4791   e-ISSN 2192-4805
ISBN 978-3-642-21451-6   e-ISBN 978-3-642-21452-3
DOI 10.1007/978-3-642-21452-3
Springer Heidelberg Dordrecht London New York

© Springer-Verlag Berlin Heidelberg 2011
This work is subject to copyright. All rights are reserved, whether the whole or part of the material is concerned, specifically the rights of translation, reprinting, reuse of illustrations, recitation, broadcasting, reproduction on microfilm or in any other way, and storage in data banks. Duplication of this publication or parts thereof is permitted only under the provisions of the German Copyright Law of September 9, 1965, in its current version, and permission for use must always be obtained from Springer. Violations are liable to prosecution under the German Copyright Law.
The use of general descriptive names, registered names, trademarks, etc. in this publication does not imply, even in the absence of a specific statement, that such names are exempt from the relevant protective laws and regulations and therefore free for general use.

Cover design: eStudio Calamar, Berlin/Figueres

Printed on acid-free paper

Springer is part of Springer Science+Business Media (www.springer.com)

To my uncle Joaquim, who first told me how far the stars really are, and to my father, my first and best teacher

Preface

This book was written as a guide for a one-week course aimed at exceptional students in their final years of secondary education, taught in July 2005, and again in July 2010, at the Mathematics Department of Instituto Superior Técnico (Lisbon). The course was intended to provide a quick but nontrivial introduction to Einstein's general theory of relativity, in which the beauty of the interplay between geometry and physics would be apparent. Given the audience, there was the limitation of using only elementary mathematics and physics; due to the time constraints, the text was deliberately written in an abbreviated style, with all nonessential material relegated to the exercises. Only the more kinematic aspects of relativity (time dilation, geodesics, the Doppler effect) were treated; the emphasis was squarely put on the geometry, with the twin paradox (seen as the Minkowski version of the triangle inequality) as the starting point.

I therefore assume that the reader knows mathematics and physics at the level of a student in the final years of secondary education: elementary algebra and geometry, basic trigonometry (essentially the definitions of sine and cosine) and a little mechanics (velocity, acceleration, mass, force and energy). Knowledge of calculus is not assumed; the mathematically sophisticated reader will recognize approximate versions of derivatives, integrals and differential equations throughout the text, and will not have difficulty to convert the corresponding approximate arguments into rigorous proofs.

Each chapter ends with a list of about ten exercises, followed by the complete solutions. These exercises should be regarded as an integral part of the book (arguably the most important part; understanding a physical theory, more often than not, means being able to calculate its consequences). Ideally, the reader should try to solve them all, turning to the solutions for confirmation.

Although it is my hope that this book will be useful to anyone who wishes to learn Einstein's theory beyond popular science accounts, I expect that the majority of readers will be astronomy, mathematics or physics undergraduates, using it either as a textbook for an introductory course or simply to get a first idea of what general relativity is all about before deciding to move on to more advanced texts (e.g. [12–16]). Relativity instructors will find this book to be a useful source of relativity problems, some of which will be interesting for graduate students and even for professional relativists.

In writing this book I have tried to think back to the long Summer vacation when I myself had just finished secondary school, and out of boredom picked up a book about relativity. I remember that the algebra seemed daunting, and that it took me a couple of weeks (and another book) to understand the twin paradox. General relativity, with its complicated looking tensor calculus, appeared hopelessly out of reach. Although those days are long gone, I have tried my best to write the book I would have liked to read back then.

Finally, I want to thank the various colleagues who read this text, or parts of it, for their many comments and corrections. Special thanks are due to Sílvia Anjos, Pedro Girão, João Pimentel Nunes, Pedro Resende, Pedro Ferreira dos Santos and Jorge Drumond Silva.

Lisbon, July 2011   José Natário

Contents

Chapter 1
Special Relativity

In this chapter we cover the basics of Einstein's special theory of relativity. Starting with a discussion of how any motion must be described relative to some frame of reference, we define inertial frames and derive the classical formulas for changing coordinates between them, called the Galileo transformation formulas. These formulas imply that velocities add, in disagreement with the surprising experimental fact that the speed of light is the same in any inertial frame. The solution to this puzzle, revealed by Einstein, is that the naive Galileo transformation formulas are actually wrong (they only work for velocities much smaller than the speed of light), and must be replaced by the more complicated Lorentz transformation formulas. We analyze some of the counter-intuitive implications of these formulas: the relativistic addition of velocities and time dilation.

1.1 Relativity of Motion

Motion is relative. As you read this you probably think that you are not moving. To be precise, however, you should think that you are not moving *with respect to the Earth's surface*. But the Earth spins (Fig. 1.1). At the latitude of Lisbon, where I am writing this, we spin at approximately 1,300 km/h (faster than the speed of sound—see Exercise 1). Moreover, the Earth moves around the Sun, at about 30 km/s (see Exercise 2), and the Sun moves around the center of the galaxy, at about 220 km/s. Therefore I am moving at 1,300 km/h with respect to the center of the Earth, at 30 km/s with respect to the Sun, and at 220 km/s with respect to the center of the galaxy.

J. Natário, *General Relativity Without Calculus*, Undergraduate Lecture
Notes in Physics, DOI: 10.1007/978-3-642-21452-3_1,
© Springer-Verlag Berlin Heidelberg 2011

Fig. 1.1 The Earth—at rest
or in motion? (Image credit:
NASA)

1.2 Inertial Frames

To study any motion we must first choose what is called a *frame of reference*.
A frame of reference is simply a system of coordinate axes with respect to which
the coordinates of each point in space can be specified. It is often attached to a
solid object (for instance the Earth), but that is not strictly necessary.

When I said I was moving at 1,300 km/h with respect to the center of the
Earth I was not being entirely accurate. What I meant was that I am moving
with this speed in the frame whose center is the center of the Earth *but which
does not spin*. This frame is (approximately) what is called an *inertial frame*,
that is, a frame where the *law of inertia* holds: any free particle moves in a
straight line with constant speed. The frame attached to the surface of the Earth
is not inertial because of the Earth's rotation, which prevents the law of inertia
from holding exactly (this is revealed in certain experiments, as for instance the
Foucault[1] pendulum). However, this frame can be considered an inertial frame
for most purposes. In reality, the non-rotating frame centered at the center of
the Earth is also not an exact inertial frame (despite being a better approximation), because of the Earth's motion around the Sun. Progressively better
approximations are the non-rotating frame centered at the Sun and the non-
rotating frame centered at the center of the galaxy. In practice, "non-rotating"
means "non-rotating with respect to the distant stars". The suggestion that the
matter in the Universe as a whole somehow determines the inertial frames is
called the *Mach[2] principle*.

The big sixteenth century controversy between geocentrism and heliocentrism
was in part a discussion about frames. In a way both parties were right: it is as
accurate to say that the Earth moves around the Sun as saying that the Sun moves
around the Earth. In the first case we using the Sun's frame, while in the second

---

[1] Jean Foucault (1819–1868), French physicist.

[2] Ernst Mach (1838–1916), Austrian physicist and philosopher.

Fig. 1.2 Galileo
transformation

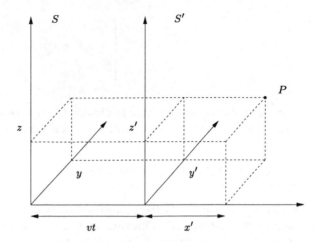

case we are adopting the frame attached to the surface of the Earth. However, since
the Sun's frame is (more approximately) inertial, it allows a much simpler
description of motion.

1.3 Galileo Transformations

If S is an inertial frame then any other frame S' with parallel axes moving with
constant velocity with respect to S is also an inertial frame. Suppose that S' moves
along the x-axis with constant velocity v, and that the two frames coincide at time
$t = 0$. If at time t a given point P has coordinates (x, y, z) in S, then it should be
clear from Fig. 1.2 that its coordinates (x', y', z') in S' are

$$\begin{cases} x' = x - vt \\ y' = y \\ z' = z \end{cases}$$

We should add to these equations

$$t' = t,$$

that is: the time measured in S' is the same as the time measured in S. This seems
so obvious that it is almost not worth writing down (and indeed for a long time no
one did). We shall soon see that it is actually wrong: as incredible as it may seem,
the time measured in the two frames is not exactly the same.

The formulas above are said to define a *Galileo*[3] *transformation* (which is
nothing more than a change of inertial frame). The inverse transformation is very
simple:

---

[3] Galileo Galilei (1564–1642), Italian astronomer, physicist and philosopher (Fig. 1.3).

$$\begin{cases} t = t' \\ x = x' + vt' \\ y = y' \\ z = z' \end{cases}$$

In other words, we just have to change the sign of v. This is what one would expect, as S is moving with respect to S' with velocity $-v$.

1.4 Velocity Addition Formula

A consequence of the Galileo transformations is the *velocity addition formula*. Suppose that the point P is moving along the x-axis. Let u be the instantaneous velocity of P in the frame S. This means that

$$u = \frac{\Delta x}{\Delta t},$$

where $\Delta x = x_2 - x_1$ is the distance travelled by P between the positions x_1 and x_2 (as measured in S) in a very small time interval $\Delta t = t_2 - t_1$ between times t_1 and t_2 (also measured in S). In S', P moves by $\Delta x'$ between positions x_1' and x_2' in the time interval $\Delta t'$ between times t_1' and t_2'. The values of t_1', t_2', x_1', x_2' are related to the values of t_1, t_2, x_1, x_2 by a Galileo transformation:

$$\begin{cases} t_1' = t_1 \\ t_2' = t_2 \\ x_1' = x_1 - vt_1 \\ x_2' = x_2 - vt_2 \end{cases}$$

Therefore

$$\begin{cases} \Delta t' = \Delta t \\ \Delta x' = \Delta x - v\Delta t \end{cases}$$

and the instantaneous velocity of P in S' is

$$u' = \frac{\Delta x'}{\Delta t'} = \frac{\Delta x - v\Delta t}{\Delta t} = \frac{\Delta x}{\Delta t} - v = u - v.$$

In other words, the velocity u of P in S is simply the sum of the velocity u' of P in S' with the velocity v of S' with respect to S.

Fig. 1.3 Galileo Galilei

1.5 Lorentz Transformations

In 1887, to their great surprise, Michelson[4] and Morley[5] (Fig. 1.4) discovered that the speed of light is the same in *all* inertial frames. This violates the velocity addition formula, according to which any object or signal is at rest in the inertial frame which moves with the same velocity.

It was Einstein[6] who first understood that this meant that the Galileo transformations formulas could not be entirely correct, and had to be replaced by the *Lorentz transformation* formulas (previously discovered by Lorentz[7] and Poincaré,[8] who had failed to correctly interpret them):

$$\begin{cases} t' = \gamma\left(t - \dfrac{vx}{c^2}\right) \\ x' = \gamma(x - vt) \end{cases}$$

where c represents the speed of light(about 300,000 km/s) and

$$\gamma = \frac{1}{\sqrt{1 - \dfrac{v^2}{c^2}}}.$$

The *special theory of relativity*, developed by Einstein in 1905, boils down to analyzing the consequences of these transformations.

---

[4] Albert Michelson (1852–1931), American physicist, winner of the Nobel prize in physics (1907).

[5] Edward Morley (1838–1923), American chemist.

[6] Albert Einstein (1879–1955), German physicist, winner of the Nobel prize in physics (1921).

[7] Hendrik Lorentz (1853–1928), Dutch physicist, winner of the Nobel prize in physics (1902).

[8] Henri Poincaré (1854–1912), French mathematician.

Fig. 1.4 Michelson, Morley, Lorentz, Poincaré and Einstein

The velocities with which we usually deal are much smaller than the speed of light, $|v| \ll c$. In this case γ is almost equal to 1, and $\frac{vx}{c^2}$ is almost equal to zero. Therefore for most applications the Lorentz transformation formulas reduce to the Galileo transformation formulas. It is only when the velocities involved become comparable to the speed of light that the Lorentz transformations become important. It is easy to check that the inverse transformation formulas are obtained (as one would expect) replacing v by $-v$ (see Exercise 6):

$$\begin{cases} t = \gamma\left(t' + \dfrac{vx'}{c^2}\right) \\ x = \gamma(x' + vt') \end{cases}$$

1.6 Relativistic Velocity Addition Formula

Notice that the Lorentz transformation formulas require $|v| < c$: *given two inertial frames, the velocity of one of them with respect to the other must be less than the speed of light.* Therefore it is never possible for a light signal to be at rest in a given inertial frame. More generally, the Lorentz transformations imply that light moves with the same speed in all inertial frames. To check this fact we need the *relativistic velocity addition formula.* Again assume that the point P is moving with instantaneous velocity u in S, travelling a distance Δx (as measured in S) in a time interval Δt (also measured in S). Then the displacement $\Delta x'$ measured in S' and the corresponding time interval $\Delta t'$ are given by

$$\begin{cases} \Delta t' = \gamma\left(\Delta t - \dfrac{v\Delta x}{c^2}\right) \\ \Delta x' = \gamma(\Delta x - v\Delta t) \end{cases}$$

Consequently, the instantaneous velocity of P in S' is

$$u' = \frac{\Delta x'}{\Delta t'} = \frac{\Delta x - v\Delta t}{\Delta t - \dfrac{v\Delta x}{c^2}} = \frac{u - v}{1 - \dfrac{uv}{c^2}}$$

In the special case when $u = c$ we obtain

$$u' = \frac{c - v}{1 - \dfrac{v}{c}} = c \cdot \frac{c - v}{c - v} = c.$$

If on the other hand $u = -c$ we have

$$u' = \frac{-c - v}{1 + \dfrac{v}{c}} = -c \cdot \frac{c + v}{c + v} = -c.$$

Therefore whenever P moves at the speed of light in S it also moves at the speed of light in S'.

1.7 Time Dilation

One of the most counter-intuitive consequences of the Lorentz transformations is the fact that the time interval between two events depends on the inertial frame in which it is measured. Suppose that an observer at rest in the inertial frame S' ($\Delta x' = 0$) measures a time interval $\Delta t'$. Then the corresponding time interval measured in the inertial frame S is

$$\Delta t = \gamma \left(\Delta t' + \frac{v \Delta x'}{c^2} \right) = \gamma \Delta t' > \Delta t'$$

(as $\gamma > 1$ whenever $v \neq 0$). This phenomenon is known as *time dilation*.

1.8 Derivation of the Lorentz Transformation Formulas

In what follows we give a derivation of the Lorentz transformation formulas, due to Einstein. Einstein started with two postulates:

1. *Relativity principle*: Any two inertial frames are equivalent.
2. *Invariance of the speed of light*: The speed of light is the same in all inertial frames.

Since the Galileo transformations are not compatible with the second postulate, we cannot expect the "obvious" formula $x' = x - vt$ to work. Suppose however that x' is *proportional* to $x - vt$, that is,

$$x' = \gamma(x - vt)$$

for some constant γ (to be determined). Since S moves with respect to S' with velocity $-v$, the first postulate requires an analogous formula for the inverse transformation:

$$x = \gamma(x' + vt').$$

Solving for t' yields

$$t' = \frac{x}{v\gamma} - \frac{x'}{v}.$$

Substituting in this formula the initial expression for x' gives

$$t' = \left(\frac{1}{\gamma} - \gamma\right)\frac{x}{v} + \gamma t.$$

We now use the second postulate. Consider a light signal propagating along the x-axis in S, passing through $x = 0$ at time $t = 0$. The position of the signal at time t will then be $x = ct$. On the other hand, the second postulate requires that the position of the signal in S' be $x' = ct'$. Therefore

$$c = \frac{x'}{t'} = \frac{\gamma(x - vt)}{\left(\frac{1}{\gamma} - \gamma\right)\frac{x}{v} + \gamma t} = \frac{xt - v}{\left(\frac{1}{\gamma^2} - 1\right)\frac{x}{vt} + 1} = \frac{c - v}{\left(\frac{1}{\gamma^2} - 1\right)\frac{c}{v} + 1},$$

implying that

$$\left(\frac{1}{\gamma^2} - 1\right)\frac{c}{v} + 1 = 1 - \frac{v}{c} \Leftrightarrow \frac{1}{\gamma^2} = 1 - \frac{v^2}{c^2} \Leftrightarrow \gamma = \pm\frac{1}{\sqrt{1 - \frac{v^2}{c^2}}}.$$

Since we must have $\gamma = 1$ for $v = 0$, we must take the positive sign. Thus

$$\frac{1}{\gamma} - \gamma = \gamma\left(\frac{1}{\gamma^2} - 1\right) = \gamma\left(1 - \frac{v^2}{c^2} - 1\right) = -\gamma\frac{v^2}{c^2},$$

and we finally obtain

$$\begin{cases} t' = \gamma\left(t - \frac{vx}{c^2}\right) \\ x' = \gamma(x - vt) \end{cases}$$

with

$$\gamma = \frac{1}{\sqrt{1 - \frac{v^2}{c^2}}}.$$

1.9 Important Formulas

- Lorentz transformations:

$$\boxed{\begin{cases} t' = \gamma\left(t - \frac{vx}{c^2}\right) \\ x' = \gamma(x - vt) \end{cases}} \quad \text{or} \quad \boxed{\begin{cases} t = \gamma\left(t' + \frac{vx'}{c^2}\right) \\ x = \gamma(x' + vt') \end{cases}} \quad \text{with} \quad \boxed{\gamma = \frac{1}{\sqrt{1 - \frac{v^2}{c^2}}}}$$

• Velocity addition (see Exercise 8):

$$u' = \frac{u - v}{1 - \dfrac{uv}{c^2}} \quad \text{or} \quad u = \frac{u' + v}{1 + \dfrac{u'v}{c^2}}$$

• Time dilation:

$$\Delta t' = \frac{\Delta t}{\gamma} = \Delta t \sqrt{1 - \frac{v^2}{c^2}}$$

1.10 Exercises

1. Show that the Earth's rotation speed at the latitude of Lisbon (about 39°) is approximately 1,300 km/h (radius of the Earth: approximately 6,400 km). Show that this speed is greater than the speed of sound (about 340 m/s).

2. Show that the speed of the Earth with respect to the Sun is about 30 km/s. (Distance from the Earth to the Sun: approximately 8.3 light-minutes).

3. A machine gun mounted in the rear of a bomber which is flying at 900 km/h is firing bullets also at 900 km/h but in the direction opposite to the flight direction. What happens to the bullets?

4. A boy throws a tennis ball at 50 km/h towards a train which is approaching at 100 km/h. Assuming that the collision is perfectly elastic, at what speed does the ball come back?

5. *Relativity of simultaneity*: The starship *Enterprise* travels at 80% of the speed of light with respect to the Earth (Fig. 1.5). Exactly midship there is a flood-light. When it is turned on, its light hits the bow and the stern at exactly the same time as seen by observers on the ship. What do observers on Earth see?

6. Check that the inverse Lorentz transformation formulas are correct.

7. *Length contraction*: Consider a ruler of length l' at rest in the frame S'. The ruler is placed along the x'-axis, so that its ends satisfy $x' = 0, x' = l'$ for all t'. Write the equations for the motion of the ruler's ends in the frame S, and show that in this frame the ruler's length is

$$l = \frac{l'}{\gamma} < l'.$$

8. Check that the relativistic velocity addition formula can be written as

$$u = \frac{u' + v}{1 + \dfrac{u'v}{c^2}}.$$

9. A rocket which is flying at 50% of the speed of light with respect to the Earth fires a missile at 50% of the speed of light (with respect to the rocket). What is the velocity of the missile with respect to the Earth when the missile is fired:

   (a) Forwards?
   (b) Backwards?

10. Two rockets fly at 50% of the speed of light with respect to the Earth, but in opposite directions. What is the velocity of one of the rockets with respect to the other?

11. The time dilation formula can be derived directly from the fact that the speed of light is the same on any inertial frame by using a *light clock* (Fig. 1.6): consider a light signal propagating in S' along the y'-axis between a laser emitter L and a detector D. If the signal travel time is $\Delta t'$ (as measured in S'), then the distance between L and D (as measured in S') is $\Delta y' = c\Delta t'$. On the other hand, as seen from S the detector D moves along the x-axis with velocity v. Therefore during the time interval Δt (as measured in S) between the emission and the detection of the signal the detector moves by $v\Delta t$ along the x-axis. Assuming that the distance Δy measured in S between the laser emitter and the detector is the same as in S', $\Delta y = \Delta y'$, derive the time dilation formula.

12. *Useful approximations*: show that if $|\varepsilon| \ll 1$ then

   (a) $\dfrac{1}{1+\varepsilon} \simeq 1 - \varepsilon$;

   (b) $\sqrt{1+\varepsilon} \simeq 1 + \dfrac{\varepsilon}{2}$, with error of order $\varepsilon^2 \ll |\varepsilon|$.

13. The time dilation formula was experimentally verified in 1971 by comparing the readings of two very precise atomic clocks. One of the clocks was kept at rest on the surface of the Earth, whereas the other was flown once around the Earth, along the parallel of latitude 39°, at an average speed of 900 km/h.

   (a) What was the difference in the readings of the two clocks? Does it make a difference whether the clock was flown eastwards or westwards?
   (b) Show that even if the moving atomic clock were transported very slowly along the parallel the two clocks would be desynchronized at the end of the journey (*Sagnac*[9] *effect*).
      (Recall that the Earth is not a perfect inertial frame because it is spinning; use the approximation $\sqrt{1 - \dfrac{v^2}{c^2}} \simeq 1 - \dfrac{v^2}{2c^2}$ for velocities v much smaller than c).

14. When cosmic rays hit the Earth's atmosphere they produce (among others) particles called *muons*, typically at an altitude of 10 km. These elementary particles are unstable, and decay in about 2.2×10^{-6} s. However, a large

---

[9] Georges Sagnac (1869–1926), French physicist.

Fig. 1.5 The *Enterprise* travels at 80% of the speed of light with respect to the Earth. (STAR TREK and related marks are trademarks of CBS Studios Inc.)

Fig. 1.6 Light clock

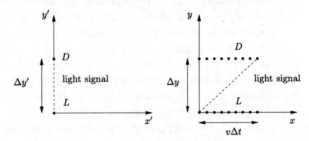

percentage of these muons is detected on the ground. What is the minimum speed at which the detected particles must be moving?

15. *Twin paradox*: Two twins, Alice and Bob, part on their twentieth birthday. While Alice remains on Earth (which is an inertial frame to a very good approximation), Bob departs at 80% of the speed of light towards Planet X, 8 light-years away from Earth. Therefore Bob reaches his destination 10 years later (as measured on the Earth's frame). After a short stay, he returns to Earth, again at 80% of the speed of light. Consequently Alice is 40 years old when she sees Bob again.

   (a) How old is Bob when they meet again?
   (b) How can the asymmetry in the twins' ages be explained? Notice that from Bob's point of view he is at rest in his spaceship and it is the Earth which moves away and then back again.

1.11 Solutions

1. The radius for the parallel through Lisbon is about $6,400 \times \cos(39°) \simeq 5,000\,\mathrm{km}$ (see Fig. 3.5), yielding a circumference of about $2\pi \times 5,000 \simeq 31,000\,\mathrm{km}$. Therefore a point at this latitude travels $31,000$ km each 24 h,

corresponding to a speed of about 1, 300 km/h. This speed is greater than the speed of sound, which is approximately $0.34 \times 3,600 \simeq 1,200$ km/h.

2. One light-minute is the distance travelled by light (whose speed is 300, 000 km/s) in one minute. The circumference of the Earth's orbit is therefore $2\pi \times 8.3 \times 60 \times 300,000$ km, and the Earth travels this distance in one year, that is, $365 \times 24 \times 3,600$ s. Dividing these two numbers yields the result.

3. By the (Galileo) velocity addition formula, the speed of the bullets with respect to the ground is $900 - 900 = 0$ km/h. Consequently, the bullets simply fall down vertically.

4. Relative to the train the ball is travelling at $50 + 100 = 150$ km/h. Since the collision is perfectly elastic, it comes back with the same speed. Relative to the boy, it then comes back at $150 + 100 = 250$ km/h. This trick is often used by space probes in a maneuver called *gravitational assist*. In this maneuver, the probe plays the role of the tennis ball and the planet plays the role of the train. Instead of colliding elastically with the planet, the probe performs a close flyby, being strongly deflected by the planet's gravitational field. By conservation of energy, this deflection behaves much like an elastic collision, allowing the probe to increase its speed considerably.

5. As seen by observers on Earth, the light hits the stern first, as the stern is moving towards the emission point (whereas the bow is moving away from the emission point). Quantitatively, let $2L$ be the *Enterprise*'s length, and assume that the floodlight is placed at $x' = 0$ and is lit at time $t' = 0$. Then its light hits the stern $(x' = -L)$ and the bow $(x' = L)$ at time $t' = \frac{L}{c}$. Since $\frac{v}{c} = 0.8$, and hence $\sqrt{1 - \frac{v^2}{c^2}} = 0.6$, the Lorentz transformation formulas tell us that in the Earth's frame the light hits the stern at time

$$t = \frac{t' + 0.8\frac{x'}{c}}{0.6} = \frac{\frac{L}{c} - 0.8\frac{L}{c}}{0.6} = \frac{L}{3c},$$

and hits the bow at time

$$t = \frac{t' + 0.8\frac{x'}{c}}{0.6} = \frac{\frac{L}{c} + 0.8\frac{L}{c}}{0.6} = \frac{3L}{c}.$$

Thus from the point of view of observers on Earth, the light takes 9 times longer to hit the bow than to hit the stern.

6. We just have to check that

$$\gamma\left(t' + \frac{vx'}{c^2}\right) = \gamma^2\left(t - \frac{vx}{c^2}\right) + \gamma^2\frac{v}{c^2}(x - vt) = \gamma^2\left(1 - \frac{v^2}{c^2}\right)t = t$$

and

$$\gamma(x' + vt') = \gamma^2(x - vt) + v\gamma^2\left(t - \frac{vx}{c^2}\right) = \gamma^2\left(1 - \frac{v^2}{c^2}\right)x = x.$$

7. The ruler's ends move according to the equations

$$x' = 0 \Leftrightarrow \gamma(x - vt) = 0 \Leftrightarrow x = vt$$

and

$$x' = l' \Leftrightarrow \gamma(x - vt) = l' \Leftrightarrow x = \frac{l'}{\gamma} + vt.$$

Therefore the ruler's length in the frame S is

$$l = \frac{l'}{\gamma} + vt - vt = \frac{l'}{\gamma} = l'\sqrt{1 - \frac{v^2}{c^2}}$$

(which is always less than l').

8. We just have to see that

$$u' = \frac{u - v}{1 - \dfrac{uv}{c^2}} \Leftrightarrow u - v = u' - \frac{u'uv}{c^2} \Leftrightarrow \left(1 + \frac{u'v}{c^2}\right)u = u' + v \Leftrightarrow u = \frac{u' + v}{1 + \dfrac{u'v}{c^2}}.$$

9. In both cases the velocity of the rocket with respect to the Earth is $v = 0.5\,c$. Therefore:

   (a) When the missile is fired forwards its velocity with respect to the rocket is $u' = 0.5\,c$. Thus its velocity with respect to the Earth will be

   $$u = \frac{u' + v}{1 + \dfrac{u'v}{c^2}} = \frac{0.5\,c + 0.5\,c}{1 + 0.5 \times 0.5} = \frac{c}{1.25} = 0.8\,c.$$

   (b) When the missile is fired backwards its velocity with respect to the rocket is $u' = -0.5\,c$. Thus its velocity with respect to the Earth will be

   $$u = \frac{u' + v}{1 + \dfrac{u'v}{c^2}} = \frac{-0.5\,c + 0.5\,c}{1 - 0.5 \times 0.5} = 0.$$

10. The frame of the rocket which is moving from left to right has velocity $v = 0.5\,c$. In the Earth's frame, the rocket which is moving from right to left has velocity $u = -0.5\,c$. With respect to the first rocket, its velocity is then

$$u' = \frac{u - v}{1 - \dfrac{uv}{c^2}} = \frac{-0.5\,c - 0.5\,c}{1 + 0.5 \times 0.5} = -\frac{c}{1.25} = -0.8\,c.$$

11. By the Pythagorean theorem, the distance traveled by the light signal in the frame S is

$$c^2\Delta t^2 = v^2\Delta t^2 + \Delta y^2 = v^2\Delta t^2 + \Delta y'^2 = v^2\Delta t^2 + c^2\Delta t'^2.$$

Solving for $\Delta t'$ we obtain

$$\Delta t' = \Delta t\sqrt{1 - \frac{v^2}{c^2}}.$$

12. If $|\varepsilon| \ll 1$ then $\varepsilon^2 \ll |\varepsilon|$ (for instance if $\varepsilon = 0.01$ then $\varepsilon^2 = 0.0001$). Therefore with an error on order ε^2 (hence negligible) we have:

(a) $(1 - \varepsilon)(1 + \varepsilon) = 1 - \varepsilon^2 \simeq 1$, whence $\dfrac{1}{1 + \varepsilon} \simeq 1 - \varepsilon$;

(b) $\left(1 + \dfrac{\varepsilon}{2}\right)^2 = 1 + \varepsilon + \dfrac{\varepsilon^2}{4} \simeq 1 + \varepsilon$, whence $\sqrt{1 + \varepsilon} \simeq 1 + \dfrac{\varepsilon}{2}$.

For instance,

(a) $\dfrac{1}{1.01} = 0.99009900... \simeq 0.99 = 1 - 0.01$:

(b) $\sqrt{1.01} = 1.00498756... \simeq 1.005 = 1 + \dfrac{0.01}{2}$.

13. In this problem the fact that the frame attached to the surface of the Earth is not an inertial frame (because of the Earth's rotation) is relevant. Consequently, we must use the inertial frame attached to the center of the Earth. As we saw in the answer to Exercise 1, the length of the parallel of latitude 39° is $L = 31,000$ km. Consequently the airplane journey took approximately $\dfrac{31,000}{900} \simeq 31.4$ h, that is, about 124,000 s.

(a) As was seen in Exercise 1, the clock on the surface of the Earth is moving at about 1,300 km/h in the frame attached to the center of the Earth. Thus when the airplane is flying eastwards it is moving at $1,300 + 900 = 2,200$ km/h in this frame, whereas when it flies westwards it is moving at $1,300 - 900 = 400$ km/h. The difference in the readings of the two clocks when the airplane has flown eastwards is then

$$124,000\left(\sqrt{1 - \frac{1,300^2}{(3,600 \times 300,000)^2}} - \sqrt{1 - \frac{2,200^2}{(3,600 \times 300,000)^2}}\right) \text{ s.}$$

Using the approximation $\sqrt{1 - \dfrac{v^2}{c^2}} \simeq 1 - \dfrac{v^2}{2c^2}$, we obtain

$$124,000 \times \frac{2,200^2 - 1,300^2}{2 \times (3,600 \times 300,000)^2} \simeq 170 \times 10^{-9} \text{ s.}$$

The difference in the readings of the two clocks when the airplane has flown westwards is

$$124,000 \left(\sqrt{1 - \frac{1,300^2}{(3,600 \times 300,000)^2}} - \sqrt{1 - \frac{400^2}{(3,600 \times 300,000)^2}} \right) \text{ s,}$$

that is, approximately

$$124,000 \times \frac{400^2 - 1,300^2}{2 \times (3,600 \times 300,000)^2} \simeq -80 \times 10^{-9} \text{ s.}$$

Thus if the clock was flown eastwards it was about 170 ns late with respect to the stationary clock, whereas if it was flown westwards it was about 80 ns early. These differences were indeed measured in the experiment, together with the corrections due to the gravitational field (see Exercise 3 in Chap. 5).

(b) Let $V \simeq 1,300$ km/h be the rotation speed of the Earth at latitude $39°$, and suppose that the clock is transported at a very small velocity v along the parallel. Then the journey duration will be $\frac{L}{v}$. If the clock is transported eastwards then the desynchronization between the fixed and the moving clock will be

$$\frac{L}{v}\sqrt{1 - \frac{V^2}{c^2}} - \frac{L}{v}\sqrt{1 - \frac{(V+v)^2}{c^2}} \simeq \frac{L}{v}\frac{(V+v)^2 - V^2}{2c^2} = \frac{L}{v}\frac{(2V+v)v}{2c^2} \simeq \frac{VL}{c^2},$$

that is,

$$\frac{\frac{1,300}{3,600} \times 31,000}{300,000^2} \simeq 120 \times 10^{-9} \text{ s.}$$

Thus the moving clock will be about 120 ns late with respect to the fixed clock. If the clock is transported westwards then the desynchronization will have the same absolute value but opposite sign, that is, the moving clock will be about 120 ns early with respect to the fixed clock.

The GPS satellite navigation system relies on ground stations which track the satellites' motions with great accuracy. These stations have very precise atomic clocks, which must be synchronized to the nanosecond. To synchronize the clocks the Sagnac effect must be taken into account.

14. If the muons are moving at a speed of v km/s, it will take them at least $\dfrac{10}{v}$ s to reach the ground. In the muons' frame, however, the elapsed time is

$$\frac{10}{v}\sqrt{1-\frac{v^2}{c^2}},$$

because of time dilation. For the muons to reach the ground this time interval must be less than $2.2 \times 10^{-6}\,\mathrm{s}$:

$$\frac{10}{v}\sqrt{1-\frac{v^2}{c^2}}<2.2\times 10^{-6} \Leftrightarrow \frac{100}{v^2}\left(1-\frac{v^2}{c^2}\right)<4.84\times 10^{-12}$$

$$\Leftrightarrow \frac{1}{v^2}-\frac{1}{c^2}<4.84\times 10^{-14} \Leftrightarrow \frac{c^2}{v^2}<1+4.4\times 10^{-3}$$

$$\Leftrightarrow \frac{v}{c} > \frac{1}{\sqrt{1+4.4\times 10^{-3}}} \simeq \frac{1}{1+2.2\times 10^{-3}}$$

$$\simeq 1-2.2\times 10^{-3}.$$

So the muons detected on the ground must be moving at more than 99.998% of the speed of light.

15. (a) Since 20 years have gone by for Alice, during which Bob was mostly moving at 80% of the speed of light, Bob must have experienced

$$20\sqrt{1-0.8^2} = 20\sqrt{0.36} = 20 \times 0.6 = 12\,\mathrm{years},$$

and so he will by 32 years old when they meet again.

  (b) The asymmetry in the twins' ages is due to the fact that only Alice is on an inertial frame, since Bob must slow down as he reaches Planet X and then speed up again to return to Earth. Although velocity is a relative concept, being an inertial observer or an accelerated observer is an absolute concept.

Chapter 2
Minkowski Geometry

In this chapter we introduce Minkowski geometry, a geometric formulation of the special theory of relativity. The starting point is the representation of events as points on the plane by means of their space and time coordinates as measured in a particular inertial frame. This is akin to identifying points on the Euclidean plane with pairs of real numbers by means of their Cartesian coordinates relative to a particular system of orthogonal axes. We show that it is possible to define a formula for the distance between two events, called the interval, which is preserved under a change of inertial frame, just as the usual formula for the Euclidean distance between two points is preserved under a change of the system of orthogonal axes. The interval, which physically is just the time measured by a free particle travelling between the two events, is very different from the Euclidean distance: the length of one side of a triangle is always larger than the sum of the lengths of the other two (twin paradox), and lines are the curves with maximum length (generalized twin paradox).

2.1 Units

Since the speed of light is the same for all observers, we may without ambiguity choose units in which $c = 1$. For instance, we can measure time in years and distances in light-years (a light-year is the distance travelled by light during 1 year, approximately 9.5×10^{12} km—see Exercise 1). Alternatively, we can measure distances in meters and time in light-meters (a light-meter is the time it takes light to travel 1 m, approximately 3.3 ns—see Exercise 1). In these units speeds do not have dimensions: they are simply given by the fraction of the speed of light which they represent.

J. Natário, *General Relativity Without Calculus*, Undergraduate Lecture
Notes in Physics, DOI: 10.1007/978-3-642-21452-3_2,
© Springer-Verlag Berlin Heidelberg 2011

Fig. 2.1 Space–time
diagram containing the
histories of: **a** a particle at
rest; **b** a particle moving with
constant velocity; **c** a light
ray; **d** an accelerating particle

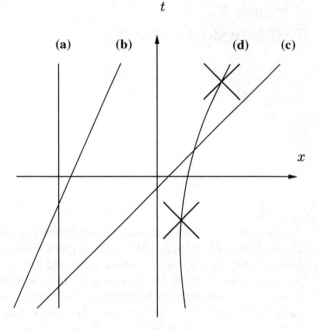

2.2 Space–Time Diagrams

To formulate the special theory of relativity geometrically we pick an inertial
frame S. Each event can be specified in this frame by giving the instant t and the
position x in which it happened. A *space–time diagram* consists in representing
events as points in the plane with Cartesian coordinates (t, x). It is traditional to
represent the coordinate t vertically.

The motion of a particle can be represented in a space–time diagram by plotting
its position x at each instant t; we thus obtain a curve, which we call the *history* of
the particle. We now consider a few examples (see Fig. 2.1):

a.   If the particle is at rest in the frame S then its position x does not change with t:
       $x = x_0$, where x_0 is a constant. Therefore the history of this particle is a vertical
       line.
b.   If the particle is moving with constant velocity v then its position at time t is
       $x = x_0 + vt$, where x_0 is a constant (representing the particle's position at
       $t = 0$). The history of this particle is then a line with slope $\frac{1}{v}$ (since its
       equation can be re-written as $t = \frac{1}{v}(x - x_0)$).
c.   A light signal moves with constant velocity $\pm c = \pm 1$, and so its position at
       time t is $x = x_0 \pm t$, where x_0 is a constant (representing the signal's position
       at $t = 0$). The history of the signal is a line with slope ± 1.

d.   If the particle moves with changing velocity then its history is a curve. As we have seen, the Lorentz transformation formulas force the particle's speed to be smaller than $c = 1$. Therefore if we imagine two light signals being emitted in opposite directions at each event in the particle's history (lines with slopes ± 1 through that event) then the particle's history cannot intersect the histories of those light signals in any other event.

2.3 Interval Between Events

To represent events on a space–time diagram we must pick an inertial frame S. Clearly the representation will change if we choose a different inertial frame S', as the coordinates (t', x') of a given event in S' are in general different from its coordinates (t, x) in S.

This situation is analogous to what happens when we introduce Cartesian coordinates in the Euclidean plane. To do so we must fix a system S of orthogonal axes. However, the choice of these axes is not unique: for instance, we can choose a different system S' of orthogonal axes which are rotated by an angle α with respect to S (Fig. 2.2). If a given point P has coordinates (x, y) in S, its coordinates (x', y') in S' are in general different. Indeed, it is not hard to show that (see Exercise 3)

$$\begin{cases} x' = x \cos \alpha + y \sin \alpha \\ y' = -x \sin \alpha + y \cos \alpha \end{cases}$$

The coordinates of point P are thus devoid of intrinsic geometric meaning, since they depend on the choice of the axes. However, the introduction of a system of orthogonal axes allows us to compute quantities with intrinsic geometric meaning, such as the distance between two points. Let P_1 and P_2 be two points with coordinates (x_1, y_1) and (x_2, y_2) in S. The coordinates of these points in S' will be (x'_1, y'_1) and (x'_2, y'_2), where

$$\begin{cases} x'_1 = x_1 \cos \alpha + y_1 \sin \alpha \\ x'_2 = x_2 \cos \alpha + y_2 \sin \alpha \\ y'_1 = -x_1 \sin \alpha + y_1 \cos \alpha \\ y'_2 = -x_2 \sin \alpha + y_2 \cos \alpha \end{cases}$$

If $\Delta x = x_2 - x_1$, $\Delta y = y_2 - y_1$, $\Delta x' = x'_2 - x'_1$ and $\Delta y' = y'_2 - y'_1$ are the differences between the coordinates of P_2 and P_1 in each of the two systems, we have

$$\begin{cases} \Delta x' = \Delta x \cos \alpha + \Delta y \sin \alpha \\ \Delta y' = -\Delta x \sin \alpha + \Delta y \cos \alpha \end{cases}$$

The distance Δs between P_1 and P_2 can be computed in S from the Pythagorean theorem:

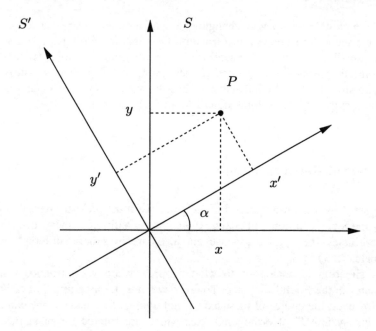

Fig. 2.2 Two systems of orthogonal axes

$$\Delta s^2 = \Delta x^2 + \Delta y^2.$$

In S' this distance is given by

$$\Delta s^2 = \Delta x'^2 + \Delta y'^2.$$

Since this distance is a geometric property, it cannot depend on the system of orthogonal axes chosen to perform the calculation. Indeed:

$$\begin{aligned}
\Delta x'^2 + \Delta y'^2 &= (\Delta x \cos\alpha + \Delta y \sin\alpha)^2 + (-\Delta x \sin\alpha + \Delta y \cos\alpha)^2 \\
&= \Delta x^2 \cos^2\alpha + \Delta y^2 \sin^2\alpha + 2\Delta x\Delta y \sin\alpha \cos\alpha + \Delta x^2 \sin^2\alpha \\
&\quad + \Delta y^2 \cos^2\alpha - 2\Delta x\Delta y \sin\alpha \cos\alpha \\
&= \Delta x^2(\sin^2\alpha + \cos^2\alpha) + \Delta y^2(\sin^2\alpha + \cos^2\alpha) \\
&= \Delta x^2 + \Delta y^2.
\end{aligned}$$

By analogy, we define the "distance" $\Delta\tau$ between two events P_1 and P_2 with coordinates (t_1, x_1) and (t_2, x_2) in S as

$$\Delta\tau^2 = \Delta t^2 - \Delta x^2,$$

where $\Delta t = t_2 - t_1$ and $\Delta x = x_2 - x_1$. Note that $\Delta\tau$ is *not* the Euclidean distance between the two events in the space–time diagram, because of the minus sign. Also note that we can only define the distance between two events such that $|\Delta x| \leq |\Delta t|$.

Pairs of events satisfying this relation are said to be *causally related*, since only in this case can one of them cause the other: the maximum speed of propagation of any signal is the speed of light, and so $\left|\frac{\Delta x}{\Delta t}\right| \leq 1$ along the history of any signal. To the "distance" $\Delta \tau$ we call the *interval* between the causally related events P_1 and P_2.

Surprisingly, the interval does not depend on the inertial frame in which it is computed:

$$\begin{aligned}
\Delta t'^2 - \Delta x'^2 &= \gamma^2(\Delta t - v\Delta x)^2 - \gamma^2(\Delta x - v\Delta t)^2 \\
&= \gamma^2(\Delta t^2 + v^2\Delta x^2 - 2v\Delta t\Delta x - \Delta x^2 - v^2\Delta t^2 + 2v\Delta t\Delta x) \\
&= \gamma^2(1 - v^2)\Delta t^2 - \gamma^2(1 - v^2)\Delta x^2 \\
&= \Delta t^2 - \Delta x^2
\end{aligned}$$

(where we used the formulas for the Lorentz transformations with $c = 1$). We can therefore regard special relativity as the study of a geometry, different from the usual Euclidean geometry, in which the distance between two points is replaced by the interval between two events. This new geometry is called *Minkowski geometry.*[1]

What is the physical meaning of the interval between two events? If $\Delta\tau \neq 0$, we have $|\Delta x| < |\Delta t|$. Therefore there exists an observer with constant velocity v which is present at both events, since

$$|v| = \left|\frac{\Delta x}{\Delta t}\right| < 1.$$

In this observer's inertial frame S', the two events happen in the same location, $\Delta x' = 0$. Therefore in this frame

$$\Delta\tau = |\Delta t'|.$$

We conclude that the interval between two events represents the time measured between them by an inertial observer which is present at both events (if nonzero). If $\Delta\tau = 0$, we have $|\Delta x| = |\Delta t|$, and hence if the interval between two events is zero then these events are placed on the history of a light signal.

2.4 Generalized Twin Paradox

The twin paradox refers to the situation illustrated by Exercise 15 in Chap. 1: Two twins, Alice and Bob, part on their 20th birthday. While Alice remains on Earth (which is an inertial frame to a very good approximation), Bob departs at 80% of the speed of light towards Planet X, 8 light-years away from Earth. Therefore Bob reaches his destination 10 years later (as measured in the Earth's frame). After a

---

[1] Hermann Minkowski (1864–1909), German mathematician Fig. 2.3.

Fig. 2.3 Hermann
Minkowski

short stay, he returns to Earth, again at 80% of the speed of light. Consequently
Alice is 40 years old when she sees Bob again. How old is Bob?

Using the time dilation formula, one can show that Bob is only 32 years old.
Let us see how to reach the same conclusion by using Minkowski geometry.
We start by picking an inertial frame. The simplest choice is the Earth's frame.
Since in this frame the Earth is at rest, its history is a vertical line, say the t-axis
($x = 0$). The history of Planet X, which is also at rest in this frame, is another
vertical line, say the line $x = 8$ (using years and light-years as units). Suppose that
Bob departs from Earth at $t = 0$, corresponding to the event O with coordinates
(0, 0). Since it takes Bob 10 years to reach Planet X (in the Earth's frame), his
arrival is the event P with coordinates (10, 8). Finally, the twins' reunion is the
event Q with coordinates (20, 0) (Fig. 2.4).

The time interval measured by Bob on the first leg of the journey is then the
interval $\overline{OP}$ between events O and P, given by

$$\overline{OP}^2 = 10^2 - 8^2 = 100 - 64 = 36,$$

that is, $\overline{OP} = 6$. The time interval measured by Bob in the return leg of the journey
is the interval $\overline{PQ}$ between events P and Q, given by

$$\overline{PQ}^2 = (20 - 10)^2 - (0 - 8)^2 = 10^2 - 8^2 = 100 - 64 = 36,$$

that is, $\overline{PQ} = 6$. Therefore the total journey takes Bob $\overline{OP} + \overline{PQ} =$
$6 + 6 = 12$ years.

The fact that Bob is younger at the twins' reunion can be geometrically
reformulated as the statement that

$$\overline{OQ} > \overline{OP} + \overline{PQ},$$

that is: the interval corresponding to the side OQ of the triangle OPQ is bigger than
the sum of the intervals corresponding to the other two sides.

Fig. 2.4 Space–time
diagram for the twin paradox

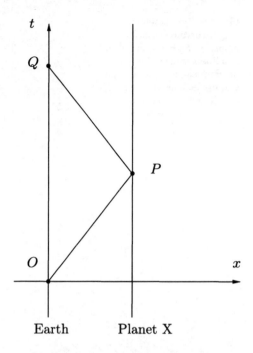

Earth   Planet X

This is exactly the opposite of what happens in Euclidean geometry, where the length of one side of a triangle is always smaller than the sum of the other two (*triangle inequality*). The triangle inequality can be used to show that the minimum length curve between two points in the Euclidean plane is a line segment. Indeed, any curve connecting two points in the plane can be approximated by a broken line joining intermediate points along the curve (Fig. 2.5). By repeatedly applying the triangle inequality it is clear that the broken line's length is bigger than the length of the line segment joining the two points. Since we can make the broken line's length as close to the curve's length as we like (by increasing the number of intermediate points), we conclude that the curve's length is necessarily bigger than the length of the line segment.

We can make the same reasoning in Minkowski geometry, using the twin paradox inequality. Since we can only compute intervals between causally related events, we can only compute "lengths" of curves such that any two points along these curves are causally related. Curves of this kind are called *causal curves*. As we saw, these are exactly the curves which represent histories of particles. The length of a causal curve must then be interpreted as the time measured by the particle along its history. If the causal curve is a line then the particle is not accelerating in any inertial frame (free particle). We can then state the *generalized twin paradox*: of all causal curves joining two given events, the one with maximum length is a line segment. Physically, of all observers who witness two given events, the one who ages the most is the one who never accelerates.

Fig. 2.5 The minimum
length curve between two
points in the Euclidean plane
is a line segment. The same
diagram can be used to show
that the maximum length
causal curve between two
events is a line segment

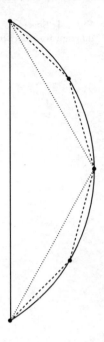

2.5 More Dimensions

For simplicity's sake, we have so far considered space–time diagrams with two
dimensions only (coordinates (t, x)). However, a complete space–time diagram has
four dimensions, corresponding to the coordinates (t, x, y, z) of events in an inertial
frame. In this case, the interval between causally related events is

$$\Delta\tau^2 = \Delta t^2 - \Delta x^2 - \Delta y^2 - \Delta z^2.$$

Unfortunately it is not easy to visualize a four dimensional space.[2] For this
reason we will use space–time diagrams with at most three dimensions, corresponding to the coordinates (t, x, y) in some inertial frame. This allows us to
consider particles and light signals which move in the (x, y)-plane. The interval
between two causally related events is in this case

$$\Delta\tau^2 = \Delta t^2 - \Delta x^2 - \Delta y^2.$$

Notice that in three dimensions the set of all points with zero interval with respect
to a given event O (that is, the set of all events along light signals through O) is now a
cone, called the *light cone* of O (Fig. 2.6). Correspondingly, the condition for a curve
to be causal is now that it be inside the light cone of each of its points.

---

[2] Mathematically, however, there is no problem in working with any number of dimensions—
even infinite dimensions.

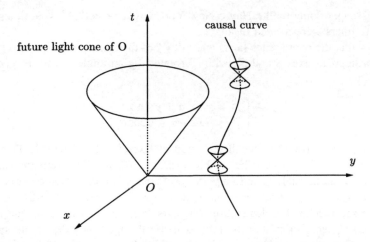

Fig. 2.6 Space–time diagram in three dimensions

2.6 Important Formulas

- Interval between events:

$$\Delta\tau^2 = \Delta t^2 - \Delta x^2$$

- Doppler effect (see Exercise 5):

$$T' = T\sqrt{\frac{1+v}{1-v}} \quad \text{with} \quad \boxed{T' = T(1+v)} \quad \text{for } |v| \ll 1$$

2.7 Exercises

1. Check that a light-year is about 9.5×10^{12} km, and that a light-meter is about 3.3×10^{-9} s.

2. *Lucas*[3] *problem*: By the end of the nineteenth century there was a regular transatlantic service between Le Havre and New York. Every day at noon (GMT) a transatlantic liner would depart from Le Havre and another one would depart from from New York. The voyage took exactly 7 days, so that the liners would also arrive at noon (GMT). Thus a liner departing from Le Havre would see a liner arriving from New York, and a liner arriving in New York would see a liner departing to Le Havre. Besides these two, how many other liners would

---

[3] Édouard Lucas (1842– 1891), French mathematician.

a passenger doing the Le Havre-New York voyage meet? At what times? How many liners were needed in total?

3. Show that the coordinates (x, y) and (x', y') of the same point P in two systems of orthogonal axes S and S', with S' rotated by an angle α with respect to S, satisfy

$$\begin{cases} x' = x \cos \alpha + y \sin \alpha \\ y' = -x \sin \alpha + y \cos \alpha \end{cases}$$

4. *Twin paradox (again)*: Recall the setup of Exercise 4 in Chap. 1: Two twins, Alice and Bob, part on their 20th birthday. While Alice remains on Earth (which is an inertial frame to a very good approximation), Bob departs at 80% of the speed of light towards Planet X, 8 light-years away from Earth. Therefore Bob reaches his destination 10 years later (as measured on the Earth's frame). After a short stay, he returns to Earth, again at 80% of the speed of light. Consequently Alice is 40 years old when she sees Bob again. Bob, however, is only 32 years old.

   a. Represent these events in the inertial frame S' in which Bob is at rest during the first leg of the journey, and check that the ages of the twins at the reunion are correct.
   b. Do the same in the inertial frame S'' in which Bob is at rest during the return leg of the journey.
   c. Imagine that each twin watches the other trough a very powerful telescope. What do they see? In particular, how much time do they experience as they see 1 year elapse for their twin?

5. *Doppler[4] effect*: Use the space–time diagram in Fig. 2.7 to show that if a light signal has period T in S then its period as measured in S' is

$$T' = T\sqrt{\frac{1 + v}{1 - v}}.$$

Moreover, show that for speeds much smaller than the speed of light, $|v| \ll 1$, this formula becomes

$$T' = T(1 + v).$$

This effect can be used to measure the velocity of an approaching (or receding) light source (e.g. a star).

6. A Klingon spy commandeers the *Einstein*, Earth's most recent starship, and escapes towards his home planet at maximum speed, 60% of the speed of light. In despair, Starfleet Command decides to install a highly experimental engine aboard the *Enterprise*, which theoretically will allow it to reach 80% of the speed of light. The installation takes 1 year, but the new engine works

---

[4] Christian Doppler (1803–1853), Austrian mathematician and physicist.

Fig. 2.7 Doppler effect

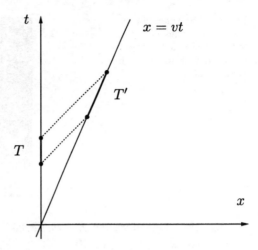

perfectly. The *Enterprise* departs in chase of the spy and captures him some time later during an exciting space battle Fig. 2.8.

a. How long does it take between the theft and the recapture of the *Einstein*:

1. According to an observer on the Earth's (inertial) frame?
2. According to the Klingon spy?
3. According to the *Enterprise* crew?

b. Starfleet has decided that if the *Enterprise* fails to recapture the *Einstein* then a radio signal will be emitted from Earth activating a secret self-destruct device aboard the *Einstein*. How long after the theft can the engineers on Earth expect to know whether to emit the radio signal?

c. When could the engineers expect confirmation of the *Einstein*'s destruction if the signal was emitted?

7. *The faster-than-light missile*[5]: During a surveillance mission on the home planet of the devious Klingons, the *Enterprise* discovers that they are preparing to build a faster-than-light missile to attack the peaceful planet Organia, 12 light-years away. Alarmed, Captain Kirk orders the *Enterprise* to depart at its maximum speed ($\frac{12}{13}$ of the speed of light) to the threatened planet. At the same time a radio signal is sent to warn the Organians about the imminent attack. Unfortunately, it is too late: 11 years later (in the frame of both planets) the Klingons finish the missile and launch it at 12 times the speed of light. So the warning reaches Organia at the same time as the missile, 12 years after being emitted, and the *Enterprise* arrives at the planet's debris 1 year later.

---

[5] This exercise is based on an exercise in [9].

Fig. 2.8 The *Enterprise*
captures the Klingon spy.
(STAR TREK and related
marks are trademarks of CBS
Studios Inc.)

    a.   How long does the *Enterprise* take to get to Organia according to its
crew?

    b.   In the planets' frame, using years and light-years as units of time and
length, let (0, 0) be the (t, x) coordinates of the event in which the
Enterprise uncovers the plot, (11, 0) the coordinates of the missile launch,
(12, 12) the coordinates of the destruction of Organia and (13, 12) the
coordinates of the arrival of the *Enterprise* at the debris. Compute the
coordinates (t', x') of the same events in the *Enterprise*'s frame.

    c.   Plot the histories of the *Enterprise*, the planets, the radio warning and the
missile in the *Enterprise*'s frame. How do events unfold in this frame?

2.8 Solutions

1. A light-year is about

$$365 \times 24 \times 3,600 \times 300,000 \simeq 9.5 \times 10^{12}\ \text{km}.$$

A light-meter is about

$$\frac{0.001}{300,000} \simeq 3.3 \times 10^{-9}\ \text{s},$$

that is, about 3.3 ns.

2. The solution of the problem becomes trivial when one draws the histories of the
liners on a space–time diagram (Fig. 2.9). Thus the passenger would meet 13
other liners, at noon and at midnight. Assuming that each liner would need
1 day to unload and reload, the service could be assured by 16 liners.

3. By elementary trigonometry, it is easy to see from Fig. 2.2 that

$$x' = \frac{x}{\cos \alpha} + \frac{y'}{\cos \alpha} \sin \alpha \Leftrightarrow x = x' \cos \alpha - y' \sin \alpha$$

Fig. 2.9 Space–time
diagram for the Lucas
problem

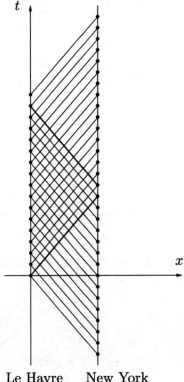

Le Havre   New York

and

$$y = \frac{y'}{\cos \alpha} + \frac{x}{\cos \alpha} \sin \alpha \Leftrightarrow y' = y \cos \alpha - x \sin \alpha.$$

Substituting the second equation in the first yields

$$x = x' \cos \alpha - y \cos \alpha \sin \alpha + x \sin^2 \alpha \Leftrightarrow x' \cos \alpha = x \cos^2 \alpha + y \cos \alpha \sin \alpha,$$

whence

$$x' = x \cos \alpha + y \sin \alpha.$$

4. Let $(0, 0)$ be the coordinates of the event O corresponding to Bob's departure in
the Earth's frame S. Then the event P in which Bob arrives on Planet X has
coordinates $(10, 8)$ and the event Q in which Bob arrives on Earth has coordinates $(20, 0)$.

   a. The inertial frame S' in which Bob is at rest during the first leg of the
       journey is moving with velocity $v = 0.8$ with respect to S. Therefore
       $\sqrt{1 - v^2} = 0.6$, and so the coordinates (t', x') of an event in S' are related
       to the coordinates (t, x) of an event in S by the Lorentz transformation
       formulas

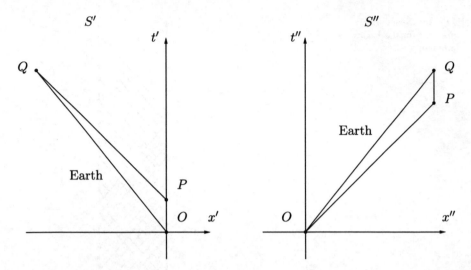

Fig. 2.10 Space–time diagrams for the twin paradox in the frames S' and S''

$$t' = \frac{t - vx}{\sqrt{1 - v^2}} = \frac{t - 0.8x}{0.6} \quad \text{and} \quad x' = \frac{x - vt}{\sqrt{1 - v^2}} = \frac{x - 0.8t}{0.6}.$$

Thus in S' event O has coordinates $(0, 0)$, event P has coordinates $(6, 0)$ (as it should), and event Q has coordinates $\left(\frac{100}{3}, -\frac{80}{3}\right)$. These events are plotted in Fig. 2.10. In we compute in this frame the time measured by Alice between events O and Q we obtain

$$\sqrt{\left(\frac{100}{3}\right)^2 - \left(\frac{80}{3}\right)^2} = \sqrt{400} = 20 \, \text{years}.$$

In the same way, the time measured by Bob between events O and P is clearly 6 years, and the time measured by Bob between events P and Q is

$$\sqrt{\left(\frac{100}{3} - 6\right)^2 - \left(\frac{80}{3}\right)^2} = \sqrt{36} = 6 \, \text{years}.$$

b. The inertial frame S'' in which Bob is at rest in the return leg of the journey is moving with velocity $v = -0.8$ with respect to S. Consequently, $\sqrt{1 - v^2} = 0.6$, and so the coordinates (t'', x'') of an event in S'' are related to the coordinates (t, x) of an event in S by the Lorentz transformation formulas

$$t'' = \frac{t - vx}{\sqrt{1 - v^2}} = \frac{t + 0.8x}{0.6} \quad \text{and} \quad x'' = \frac{x - vt}{\sqrt{1 - v^2}} = \frac{x + 0.8t}{0.6}.$$

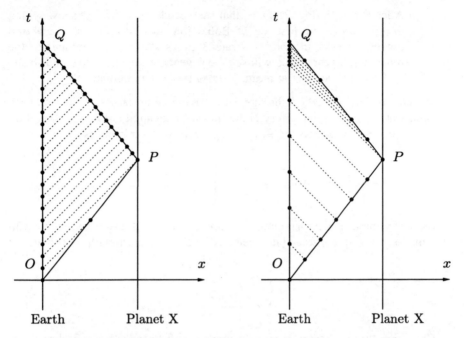

Fig. 2.11 Space–time diagram for the twin paradox

Therefore in S'' event O has coordinates $(0, 0)$, event P has coordinates $\left(\frac{82}{3}, \frac{80}{3}\right)$, and event Q has coordinates $\left(\frac{100}{3}, \frac{80}{3}\right)$. These events are plotted in Fig. 2.10. If we compute in this frame the time measured by Alice between events O and Q we obtain again

$$\sqrt{\left(\frac{100}{3}\right)^2 - \left(\frac{80}{3}\right)^2} = \sqrt{400} = 20 \, \text{years}.$$

In the same way, the time measured by Bob between events O and P is

$$\sqrt{\left(\frac{82}{3}\right)^2 - \left(\frac{80}{3}\right)^2} = \sqrt{36} = 6 \, \text{years}.$$

Finally, the time measured by Bob between events P and Q is clearly $\frac{100}{3} - \frac{82}{3} = \frac{18}{3} = 6 \, \text{years}.$

c. At event P Bob is receiving light which left Earth in $t = 2$, meaning that in the 6 years of the first leg of the journey Bob saw only 2 years of Alice's life (Fig. 2.11). Consequently, in the first leg of the journey Bob saw Alice moving in slow motion, at a rate 3 times slower than normal. In the 6 years of the return leg, Bob will see the remaining 18 years which Alice will experience, and so he will see her moving at a rate 3 times faster than normal. On the other hand, light emitted at event P reaches

Alice at $t = 18$ (Fig. 2.11), so that she spends the first 18 years watching the 6 years of the first leg of Bob's journey. Therefore she saw Bob moving in slow motion, at a rate 3 times slower than normal. In the remaining 2 years Alice will see the 6 years of the return leg, and so she will see Bob moving at a rate 3 times faster than normal.

5. We can model a periodic light signal as a sequence of flashes. If the first flash occurs at time $t = t_0$, its history is the line with equation $t = t_0 + x$. Therefore the observer in S' detects the flash in the event with coordinates

$$\begin{cases} t = t_0 + x \\ x = vt \end{cases} \quad \Leftrightarrow \quad \begin{cases} t = \dfrac{t_0}{1 - v} \\ x = \dfrac{vt_0}{1 - v} \end{cases}$$

Similarly, the second flash is emitted at time $t = t_0 + T$, its history is the line with equation $t = t_0 + T + x$, and it is detected in S' in the event with coordinates

$$\begin{cases} t = \dfrac{t_0 + T}{1 - v} \\ x = \dfrac{v(t_0 + T)}{1 - v} \end{cases}$$

Consequently, the time interval measured in S' between the two flashes is

$$T' = \sqrt{\left(\frac{t_0 + T}{1 - v} - \frac{t_0}{1 - v}\right)^2 - \left(\frac{v(t_0 + T)}{1 - v} - \frac{vt_0}{1 - v}\right)^2} = \sqrt{\frac{T^2}{(1 - v)^2} - \frac{v^2 T^2}{(1 - v)^2}}$$

$$= T\sqrt{\frac{1 - v^2}{(1 - v)^2}} = T\sqrt{\frac{(1 - v)(1 + v)}{(1 - v)^2}} = T\sqrt{\frac{1 + v}{1 - v}}.$$

If $v = 0.8$, say, we have

$$\sqrt{\frac{1 + v}{1 - v}} = \sqrt{\frac{1.8}{0.2}} = \sqrt{9} = 3.$$

This is consistent with what the twins in Exercise 4 see through their telescopes during the first leg of Bob's journey: each year for the faraway twin is observed during a 3-year period. If, on the other hand, $v = -0.8$, we have

$$\sqrt{\frac{1 + v}{1 - v}} = \sqrt{\frac{0.2}{1.8}} = \sqrt{\frac{1}{9}} = \frac{1}{3}.$$

Indeed, during the return leg of the journey each twin observes a 3-year period for the faraway twin per year.

For $|v| \ll 1$ we can apply the formula

$$\frac{1}{1 - v} \simeq 1 + v$$

to obtain the approximation

$$T' = T\sqrt{\frac{1+v}{1-v}} \simeq T\sqrt{(1+v)^2} = T(1+v).$$

6. a. 1. We assume that the theft of the *Einstein* is the event with coordinates
   $(0, 0)$ in the Earth's frame S. Then the history of the *Einstein* from that
   event on is represented by the line $x = 0.6t$. The *Enterprise* leaves Earth
   in the event $(1, 0)$, and its history from that event on is the line $x =
   0.8(t - 1)$. Therefore the *Enterprise* reaches the *Einstein* in the event
   with coordinates

   $$\begin{cases} x = 0.6t \\ x = 0.8(t-1) \end{cases} \Leftrightarrow \begin{cases} x = 0.6t \\ 0.2t = 0.8 \end{cases} \Leftrightarrow \begin{cases} t = 4 \\ x = 2.4 \end{cases}$$

   From the point of view of an observer in the Earth's inertial frame, the
   Einstein is then captured 4 years after the theft.
   2. According to the Klingon spy, the time interval between commandeering
   the *Einstein* and being captured is

   $$\sqrt{4^2 - 2.4^2} = 3.2 \, \text{years}.$$

   3. According to the *Enterprise* crew, the chase takes

   $$\sqrt{(4 - 1)^2 - 2.4^2} = 1.8 \, \text{years},$$

   and so they experience $1 + 1.8 = 2.8$ years between the theft and the
   recapture of the *Einstein*.
   b.   Since the information about the outcome of the battle will propagate at
   most at the speed of light, only after $4 + 2.4 = 6.4$ years can the engineers expect to know whether to send the self-destruct signal.
   c.   In the worst case the *Einstein* keeps moving away from the Earth at 60%
   of the speed of light, corresponding to the history $x = 0.6t$. The history of
   the self-destruct signal is the line $t = 6.4 + x$. Consequently, the *Einstein*'s self-destruction would occurs at the event with coordinates

   $$\begin{cases} x = 0.6t \\ t = 6.4 + x \end{cases} \Leftrightarrow \begin{cases} x = 9.6 \\ t = 16 \end{cases}$$

   Light from this event would reach the Earth at $t = 16 + 9.6 = 25.6 \, \text{years}$.
   So the engineers could expect confirmation of the *Einstein*'s self-
   destruction within 25.6 years after the theft.
7. In the planets' frame S, let $(0, 0)$ be the coordinates of the event O where the
   plot is uncovered, $(11, 0)$ the coordinates of the event L where the missile is
   launched, $(12, 12)$ the coordinates of the event D in which Organia is destroyed,

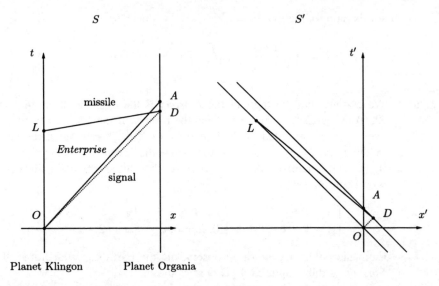

Fig. 2.12 Space–time diagrams for the faster-than-light missile

and (13, 12) the coordinates of the event A where the *Enterprise* arrives at the debris. These events are represented in Fig. 2.12.

a. The duration of the journey for the *Enterprise* crew is simply the interval $\overline{OA}$, that is,

$$\sqrt{13^2 - 12^2} = \sqrt{25} = 5 \text{ years.}$$

b. The *Enterprise*'s frame S' moves with velocity $v = \frac{12}{13}$ with respect to S. Consequently, $\sqrt{1 - v^2} = \frac{5}{13}$, and thus the coordinates (t', x') of an event in S' are related to the coordinates (t, x) of an event in S by the Lorentz transformations

$$t' = \frac{t - vx}{\sqrt{1 - v^2}} = \frac{13t - 12x}{5} \quad \text{and} \quad x' = \frac{x - vt}{\sqrt{1 - v^2}} = \frac{13x - 12t}{5}.$$

Thus in S' event O has coordinates $(0, 0)$, event L has coordinates $(28.6, -26.4)$, event D has coordinates $(2.4, 2.4)$, and event A has coordinates $(5, 0)$ (as it should).

c. These events are plotted in the *Enterprise*'s frame S' in Fig. 2.12. In S' the sequence of events is surreal: planet Organia explodes for no reason; the faster-than-light missile jumps from the debris and flies backwards towards planet Klingon, where an exact replica is being built; finally, the two missiles disappear simultaneously at event L. This illustrates the kind of absurd situations which can occur if faster-than-light speeds are permitted.

Chapter 3
Non-Euclidean Geometry

In this chapter we discuss the non-Euclidean geometry of curved surfaces, using the sphere as our primary example. We find that all the information about the geometry of the surface is contained in the expression for the distance between two nearby points in some coordinate system, called the metric. For example, the distance between two distant points can be found from the metric by determining and measuring the minimum length curve (geodesic) which connects them. Geodesics take up the role played by straight lines in the Euclidean geometry of the plane, and lead to strange new geometries, where the inner angles of a triangle do not necessarily add up to π. By considering small triangles, we define a number at each point of the surface, called the curvature, which measures how much the geometry at that point deviates from being Euclidean. This concept, as we shall see, plays an important role in Einstein's general theory of relativity.

3.1 Curvilinear Coordinates

Besides the usual Cartesian coordinates, there are many other possible choices of coordinates in the plane (traditionally called *curvilinear coordinates*). An example that naturally occurs in many situations (e.g. radars) are the so-called *polar coordinates* (r, θ), in which each point is identified by its distance r to the origin and by the angle θ between its position vector and the x-axis (Fig. 3.1).

As already seen, the distance Δs between two points P_1 and P_2 in the plane is given in Cartesian coordinates by

$$\Delta s^2 = \Delta x^2 + \Delta y^2.$$

How can one compute distances in polar coordinates? We start by observing that fixing θ and varying r by Δr corresponds to travelling a distance Δr along a ray through the origin (Fig. 3.2). Similarly, fixing r and varying θ by $\Delta\theta$ corresponds to travelling a distance $r\Delta\theta$ along the circle with center at the origin and

J. Natário, *General Relativity Without Calculus*, Undergraduate Lecture Notes in Physics, DOI: 10.1007/978-3-642-21452-3_3,
© Springer-Verlag Berlin Heidelberg 2011

Fig. 3.1 Polar coordinates

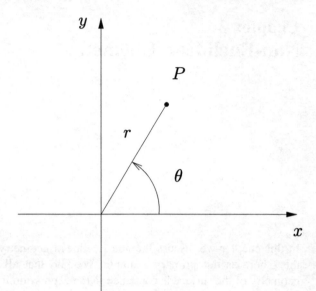

Fig. 3.2 Distance in polar coordinates

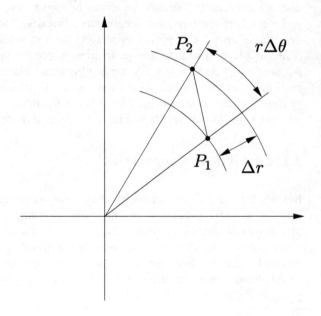

radius r. Since the ray is perpendicular to the circle, the Pythagorean theorem tells us that if two points are very close (so that the bit of circle is almost straight) then the distance between them is approximately

$$\Delta s^2 = \Delta r^2 + r^2 \Delta \theta^2$$

(with increasing accuracy as the points become closer).

We can use this formula to compute the length of a curve in polar coordinates approximating it by a broken line. The distance between two points which are not close can then be computed by computing the length of the line segment connecting them.

3.2 The Sphere

Euclidean geometry describes the geometry of the plane. However, it is often useful to understand the geometry of other surfaces. For instance, to plan a long sea or air voyage it is necessary to understand the geometry of the sphere.

Just as in the plane, the first thing to do is to choose a system of coordinates. To do so, we fix a great circle which we call the *equator*. Recall that a *great circle* is simply the intersection of the sphere with a plane through its center. In the case of the Earth's surface, the natural choice for equator is the great circle defined by the plane perpendicular to the rotation axis. The circles obtained by intersecting the sphere with planes parallel to the equator's plane are called *parallels*. Note that the parallels are not great circles. The points of intersection of the sphere with the line through the center of the sphere which is perpendicular to the equator's plane are called the *poles*. The great circle arcs between the poles are called the *meridians*. Among these we choose one which we call the *principal meridian*. Unlike the equator, there is no natural choice of principal meridian in the case of the Earth, the convention being to choose the meridian through the Greenwich astronomical observatory (near London). Any meridian can be identified by the angle φ with the principal meridian. To this angle we call the meridian's *longitude*. Similarly, any parallel can be identified by the angle θ with the equator, measured along any meridian. To this angle we call the parallel's *latitude*. Any point P on the sphere's surface can be specified by indicating the parallel and the meridian to which it belongs (Fig. 3.3). We can then use (θ, φ) as coordinates on the surface of the sphere.

Taking $-\frac{\pi}{2} \leq \theta \leq \frac{\pi}{2}$ and $-\pi \leq \varphi \leq \pi$, we can then represent the sphere as a rectangle in the plane (Fig. 3.4). In this representation, the horizontal axis corresponds to the equator, the vertical axis corresponds to the principal meridian and, more generally, horizontal lines represent parallels and vertical lines represent meridians. We then say that the rectangle is a map (or a chart) for the sphere.

Since the sphere is very different from a rectangle, it is no big surprise that the map given by the coordinates (θ, φ) is not entirely accurate, representing certain points more than once. Namely, the lines $\varphi = -\pi$ and $\varphi = \pi$ represent the same meridian, whereas the lines $\theta = \frac{\pi}{2}$ and $\theta = -\frac{\pi}{2}$ correspond each to a single point (each of the poles). Moreover, this map distorts distances: for instance, all parallels are represented by line segments with the same length, when in fact they have different lengths. More precisely, if R is the radius of the sphere then it is easy to see that the radius of the parallel of latitude θ is $R \cos \theta$ (Fig. 3.5), and its length is therefore $2\pi R \cos \theta$.

Fig. 3.3 Coordinates on the
sphere

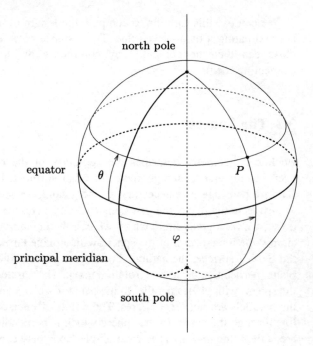

Fig. 3.4 Map of the sphere

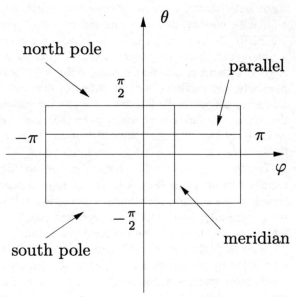

To compute distances on the sphere we start by observing that fixing φ and
varying θ by $\Delta\theta$ corresponds to travelling a distance $R\Delta\theta$ along the meridian of
longitude φ (Fig. 3.6). Similarly, fixing θ and varying φ by $\Delta\varphi$ corresponds to
travelling a distance $R\cos\theta\Delta\varphi$ along the parallel of latitude θ. Since the meridians

Fig. 3.5 Radius of a parallel

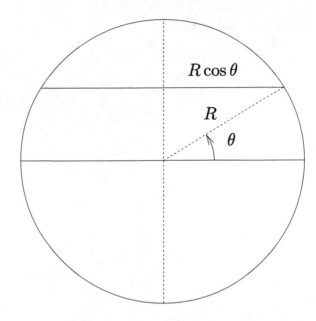

are orthogonal to the parallels, the Pythagorean theorem tells us that if the points are very close (so that the curve segments are approximately straight) then the distance between them is approximately

$$\Delta s^2 = R^2 \Delta\theta^2 + R^2 \cos^2\theta \Delta\varphi^2$$

(with increasing accuracy as the points become closer).

The expression for the distance Δs^2 between two nearby points on a given surface is called the *metric* of that surface. For instance, we have already seen that the metric of the plane is written $\Delta s^2 = \Delta x^2 + \Delta y^2$ in Cartesian coordinates, and $\Delta s^2 = \Delta r^2 + r^2\Delta\theta^2$ in polar coordinates. It is important to stress that although these two expressions are different, they contain the same information: the distance between two nearby points will have the same value whether it is computed in Cartesian or in polar coordinates.

3.3 Geodesics

Now that we have the expression for the metric of a sphere of radius R, we would like to be able to compute the distance between two points on the sphere (not necessarily close). Recall that in the plane this is done by computing the length of the line segment (that is, the curve with minimum length) joining the two points. Thus we need to identify the sphere's geodesics (that is, the minimum length curves). These curves will play the same role in the geometry of the sphere as straight lines do in the geometry of the plane.

Fig. 3.6 Distance on the
sphere

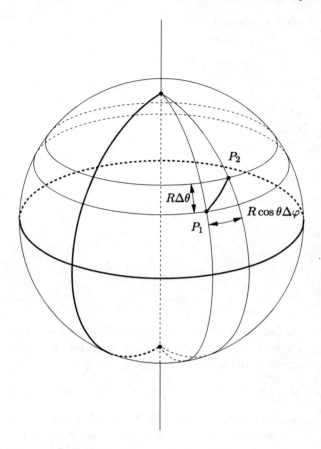

Let P and Q be two points on the sphere. It is always possible to choose
coordinates (θ, φ) such that P and Q are on the meridian $\varphi = 0$. Consider an
arbitrary curve connecting P to Q. We approximate this curve by a broken line by
choosing nearby points

$$P_0 = P, P_1, P_2, \ldots, P_{N-1}, P_N = Q$$

along the curve. These points have coordinates

$$(\theta_0, 0), (\theta_1, \varphi_1), (\theta_2, \varphi_2), \ldots, (\theta_{N-1}, \varphi_{N-1}), (\theta_N, 0).$$

If Δs_i is the distance between P_{i-1} and P_i (with $i = 1, 2, \ldots, N$) then the curve's
length is

$$l = \Delta s_1 + \Delta s_2 + \cdots + \Delta s_N.$$

According to the expression for the metric, we have approximately

$$\Delta s_i^2 = R^2(\theta_i - \theta_{i-1})^2 + R^2 \cos^2 \theta_i (\varphi_i - \varphi_{i-1})^2,$$

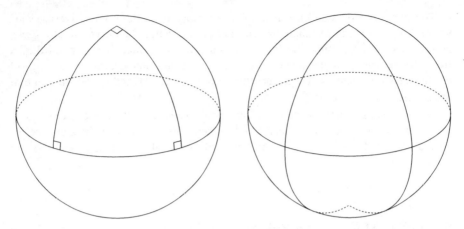

Fig. 3.7 Spherical triangle with three right angles and polygon with two sides

and therefore

$$\Delta s_i \geq R(\theta_i - \theta_{i-1}).$$

We conclude that

$$l \geq R(\theta_1 - \theta_0 + \theta_2 - \theta_1 + \cdots + \theta_N - \theta_{N-1}) = R(\theta_N - \theta_0),$$

which is exactly the length of the meridian arc between P and Q. Hence the minimum distance between two points on the sphere is measured along the smaller great circle arc joining them. In other words, the geodesics of the sphere are its great circles.

3.4 Curvature

Having identified the sphere's geodesics, we can now begin to study the sphere's geometry, which is quite different from the geometry of the plane. For instance, the sum of the internal angles of a triangle is always bigger than π : Fig. 3.7 depicts a triangle on the sphere with three right angles.

The difference between the sum of the internal angles of a spherical triangle and π is called the *spherical excess*. For instance, the spherical excess for the the triangle in Fig. 3.7 is $\frac{\pi}{2}$. The *average curvature* of a spherical triangle is the ratio between the spherical excess and the triangle's area. Recalling that the area of a sphere of radius R is $4\pi R^2$, and noting that the complete sphere is formed by 8 triangles as in Fig. 3.7, we conclude that the average curvature of that triangle is

$$\frac{\pi/2}{4\pi R^2/8} = \frac{1}{R^2}.$$

The *curvature* of a surface at a point is simply the value of the average curvature for a very small triangle around that point. Hence the curvature at a given point measures how much the local geometry of the surface at that point differs from the geometry of the plane. The sphere is a constant curvature surface: all triangles have the same average curvature (see Exercise 4), and so the curvature of the sphere at any point is $\frac{1}{R^2}$.

An interesting consequence of the spherical excess is that there exist polygons on the sphere with only two sides (for instance polygons whose sides are meridians, as depicted in Fig. 3.7). The existence of these polygons is thus a sign of the presence of curvature.

3.5 Other Maps of the Sphere

Just like in the plane, there are many possible choices of coordinates on the sphere. One possibility is to use the so-called *cylindrical projection*, which consists on projecting each point P of the sphere to a point Q in the cylindrical surface which is tangent to the sphere at the equator, perpendicularly from the axis (Fig. 3.8). The cylindrical surface can then be unrolled into a rectangle with height $2R$ and length $2\pi R$, on which we choose Cartesian coordinates (x, y) so that the equator projects to the x-axis, the parallels to horizontal lines and the meridians to vertical lines. Similarly to what we have done for the coordinates (θ, φ), one can show that the sphere's metric in these coordinates is

$$\Delta s^2 = \left(1 - \frac{y^2}{R^2}\right)\Delta x^2 + \left(1 - \frac{y^2}{R^2}\right)^{-1}\Delta y^2.$$

This expression has the particularity that the coefficients of Δx^2 and Δy^2 are inverse of each other. One can show that this happens if and only if the projection preserves areas, that is, the area of a given figure on the sphere is equal to the area of its representation on the map.

Another famous projection is the so-called *stereographic projection*, which projects each point P of the sphere to a point Q on the plane containing the equator from the north pole (Fig. 3.9). Choosing Cartesian coordinates (x, y) on the plane with origin at the center of the sphere, we see that the equator projects to itself (that is, a circle of radius R and center at the origin), the parallels to concentric circles and the meridians to lines through the origin. The sphere's metric in these coordinates is

$$\Delta s^2 = 4\left(1 + \frac{x^2}{R^2} + \frac{y^2}{R^2}\right)^{-2}\Delta x^2 + 4\left(1 + \frac{x^2}{R^2} + \frac{y^2}{R^2}\right)^{-2}\Delta y^2.$$

This expression has the particularity that the coefficients of Δx^2 and Δy^2 are equal. One can show that this happens if and only if the projection preserves angles, that is,

Fig. 3.8 Cylindrical
projection

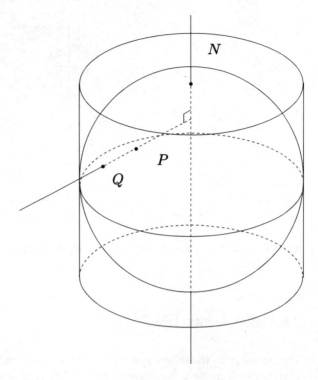

Fig. 3.9 Stereographic
projection

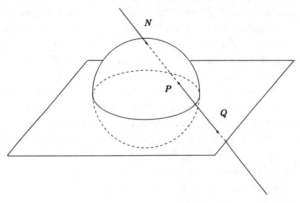

the angle between two intersecting curves on the sphere is equal to the angle between
its representations on the map (which is then called a *conformal map*).

The map that uses latitude and longitude as coordinates does not preserve areas
nor is it conformal. The most common world maps use the so-called *Mercator*[1]
projection, which can be obtained from this map (or the cylindrical projection
map) by deforming it vertically until it becomes conformal. This has the advantage

---

[1] Gerardus Mercator (1512–1594), Flemish cartographer.

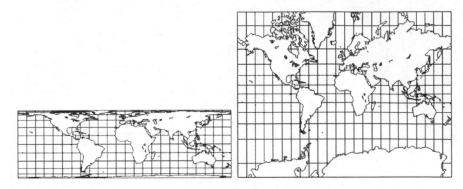

Fig. 3.10 World maps using the cylindrical and Mercator projections. (Image credit: faculty of geo-information science and earth observation, University of Twente, The Netherlands)

that the curves of constant heading are represented by straight lines on the map (Fig. 3.10).

3.6 Other Geometries

The geometry of the sphere is one example of a non-Euclidean geometry. For each different surface we will in general have a different geometry, most of which will not have constant curvature. Many will have negative curvature, that is, the sum of a triangle's internal angles will be smaller than π. Others will be flat, that is, the sum of a triangle's internal angles will be equal to π. Examples of the latter are, surprisingly, the cylinder and the cone. That is why we can roll a sheet of paper into a cylinder or a cone (but not a sphere).

By analogy, we can consider curved spaces with three or more dimensions. Although it is in general impossible to visualize such spaces, all we need to study their geometry is the expression of the metric in some coordinate system.

The ideas of non-Euclidean geometry were mostly developed by Lobachevsky,[2] Bolyai,[3] Gauss[4] and Riemann (Fig. 3.11),[5] who generalized them to spaces with any number of dimensions. For this reason, the metric of a surface (or more generally of a curved space with more dimensions) is usually called a *Riemannian metric*.

---

[2] Nikolai Ivanovich Lobachevsky (1792–1856), Russian mathematician.

[3] János Bolyai (1802–1860), Hungarian mathematician.

[4] Carl Friedrich Gauss (1777–1855), German mathematician and astronomer.

[5] Georg Bernhard Riemann (1826–1866), German mathematician.

Fig. 3.11 Lobachevsky, Bolyai, Gauss and Riemann

3.7 Exercises

1. Lisbon and New York are approximately at the same latitude (40°). However, a plane flying from Lisbon to New York does not depart from Lisbon heading west. Why not?
2. Consider two points in the sphere at the same latitude $\theta = \frac{\pi}{4}$ but with longitudes differing by π. Show that the distance between these two points is smaller than the length of the parallel arc between them.
3. Decide whether the following statements are true of false for the plane's geometry and for the sphere's geometry:

   a. All geodesics intersect.
   b. If two distinct geodesics intersect, they do so at a unique point.
   c. Given a geodesic γ and a point $p \notin \gamma$, there exists a unique geodesic γ' containing p which does not intersect γ (Euclid's' fifth postulate).
   d. There exists a geodesic containing any two distinct points.
   e. There exists a unique geodesic containing any two distinct points.
   f. There exist points arbitrarily far apart.
   g. All geodesics are closed curves.
   h. The sum on a triangle's internal angles is π.
   i. The length of a circle with radius r is $2\pi r$.
   j. The area enclosed by a circle of radius r is πr^2.

4. Show that the average curvature of a spherical triangle in which one of the vertices is the north pole and the other two vertices are on the equator is $\frac{1}{R^2}$.
5. What is the relation between the sum of the internal angles of a two-sided polygon on the sphere and its area?
6. Check that the cylindrical projection preserves the area of spherical triangles with one vertex on the north pole and the other two vertices on the equator.
7. Show that neither the stereographic projection nor the sphere's map using latitude and longitude as coordinates preserve areas.
8. Show that no map of the sphere can simultaneously preserve areas and be conformal.

3.8 Solutions

1. Because the shortest distance between Lisbon and New York is not measured along the parallel through the two cities, but along the great circle which contains them. This circle is the intersection of the Earth's surface with the plane defined by Lisbon, New York and the center of the Earth. Therefore the shortest path between Lisbon and New York is a curve whose initial heading is northwest.

2. The meridians of the two points form a great circle, and the angle between them is $\frac{\pi}{2}$. The distance between the two points is then $\frac{\pi R}{2}$. The radius of the parallel through the two points is $R\cos\left(\frac{\pi}{4}\right) = \frac{\sqrt{2}R}{2}$, and the difference in latitudes is π. The length of the parallel arc between them is then $\frac{\pi\sqrt{2}R}{2} > \frac{\pi R}{2}$.

3. a. Plane: false (there exist parallel lines). Sphere: true.
    b. Plane: true. Sphere: false (they always intersect at two points).
    c. Plane: true. Sphere: false.
    d. Plane: true. Sphere: true.
    e. Plane: true. Sphere: false (for example any meridian contains the poles).
    f. Plane: true. Sphere: false (the maximum distance between two points is πR).
    g. Plane: false. Sphere: true.
    h. Plane: true. Sphere: false.
    i. Plane: true. Sphere: false (for instance the equator is a circle centered at the north pole with radius $r = \frac{\pi}{2}R$, but its length is $2\pi R = 4r \neq 2\pi r$).
    j. Plane: true. Sphere: false (for example the equator encloses an area $2\pi R^2 = \frac{8}{\pi}r^2 \neq \pi r^2$).

4. Let α be the angle at the north pole. Since the angles on the equator are right angles, the spherical excess is exactly α. On the other hand, the triangle contains a fraction $\frac{\alpha}{2\pi}$ of the northern hemisphere's area $2\pi R^2$, that is, αR^2. We conclude that the triangle's average curvature is

$$\frac{\alpha}{\alpha R^2} = \frac{1}{R^2}.$$

5. Let α be the common value of the polygon's internal angles. Then it contains a fraction $\frac{\alpha}{2\pi}$ of the sphere's area $4\pi R^2$, that is, $2\alpha R^2$. We conclude that the ratio between the sum of the polygon's internal angles and its area is

$$\frac{2\alpha}{2\alpha R^2} = \frac{1}{R^2}.$$

6. If α is the angle at the north pole, then the triangle contains a fraction $\frac{\alpha}{2\pi}$ of the northern hemisphere's area $2\pi R^2$, that is, αR^2. It is easy to see that the triangle projects to a rectangle with the same base length αR and height equal to the sphere's radius R. Therefore, the area of the projection is also αR^2.

7. In the case of the stereographic projection, the product of the coefficients of Δx^2 and Δy^2 is

$$16\left(1 + \frac{x^2}{R^2} + \frac{y^2}{R^2}\right)^{-4} \neq 1.$$

For the sphere's map using latitude and longitude as coordinates, the product of the coefficients of $\Delta\theta^2$ and $\Delta\varphi^2$ is

$$R^4 \cos^2 \theta \neq 1.$$

8. A conformal map of the sphere is of the form

$$\Delta s^2 = A(x,y)\Delta x^2 + A(x,y)\Delta y^2$$

for some coefficient $A(x,y)$. If it preserved areas, we would have

$$A^2 = 1 \Rightarrow A = 1.$$

But then the sphere's metric in this coordinate system would be

$$\Delta s^2 = \Delta x^2 + \Delta y^2,$$

which is the metric of the Euclidean plane in Cartesian coordinates. In particular, the sphere's geodesics would be represented by straight lines on the map, and the sum of a triangle's internal angles would always be π, which we know not to be true for the sphere. Thus no conformal map of the sphere can preserve areas.

Chapter 4
Gravity

In this chapter we review the main ideas of Newtonian gravity. Starting with Newton's law of universal gravitation for point particles (or spherically symmetric bodies), we recall the usual expression for the gravitational potential energy, and use it to derive the formula for the escape velocity. Writing the laws of conservation of energy and angular momentum in polar coordinates, we obtain the differential equations for free-falling motion in the gravitational field of a spherically symmetric body, and explain how these equations determine the motion given initial conditions. As an example, we compute the speed of a circular orbit, and use it to estimate the conditions under which we should expect relativistic corrections to Newtonian gravity.

4.1 Newton's Law of Universal Gravitation

In Newtonian physics gravity is simply an attractive force between any two bodies along the line that connects them. If the bodies have masses M and m and are separated by a distance r then the intensity F of the force between them is given by *Newton's law of universal gravitation*:

$$F = \frac{GMm}{r^2},$$

where G is the so-called *universal gravitational constant*. Rigorously, this formula holds for point masses only; the force between two bodies with non-negligible dimensions is obtained subdividing them into very small pieces and adding up the contributions of each pair of pieces. One can show that for a *spherically symmetric body* the result is the same as if all the mass were placed at the center. In particular, the gravitational acceleration of a point particle of mass m due to a spherically symmetric body of mass M is

$$g = \frac{F}{m} = \frac{GM}{r^2},$$

J. Natário, *General Relativity Without Calculus*, Undergraduate Lecture Notes in Physics, DOI: 10.1007/978-3-642-21452-3_4, © Springer-Verlag Berlin Heidelberg 2011

where r is the distance to the center of the body. Note that g does not depend on the value of the mass m; this explains why the gravitational acceleration is the same for all objects.

One can show that the gravitational potential energy of the system formed by the masses M and m is

$$U = -\frac{GMm}{r}$$

(where by convention the potential energy of two masses infinitely far apart is zero). This energy is negative, since gravity is attractive: we have to spend energy to separate the two masses. The *gravitational potential* (gravitational potential energy per unit of mass) due to a spherically symmetric body of mass M is then

$$\phi = -\frac{GM}{r}.$$

4.2 Units

To simplify our formulas we will use *geometrized units*, in which besides $c = 1$ one also has $G = 1$. In these units masses are measured in meters. For instance, the mass of the Sun in geometrized units is about 1.5 km (see Exercise 5), whereas the mass of the Earth is about 4.5 mm (see Exercise 1). We shall see later that this means that a black hole with the mass of the Sun has a radius of $2 \times 1.5 = 3$ km, whereas a black hole with the mass of the Earth has a radius of $2 \times 4.5 = 9$ mm.

4.3 Escape Velocity

Consider the free-falling motion of a point particle of negligible mass m (for instance the Earth) in the gravitational field of a spherically symmetric body of mass M (for instance the Sun), which we can assume to be fixed at the origin of an inertial frame. One can show that the particle's *mechanical energy* (per unit mass),

$$E = \frac{1}{2}v^2 - \frac{M}{r},$$

is conserved along the motion (where v is the particle's velocity). In particular,

$$\frac{M}{r} \geq -E.$$

Consequently, if $E < 0$ then

$$r \leq -\frac{M}{E},$$

that is, m will never venture more than $-\dfrac{M}{E}$ away from the center of M. If m is to move arbitrarily far away from M we must have $E \geq 0$. This means that for m to escape the attraction of M from a point at a distance r from the center of M it must be launched with velocity

$$v \geq \sqrt{\frac{2M}{r}}.$$

This minimum velocity is called the *escape velocity*.

4.4 Kepler Laws

It is easy to see that the motion of the mass m happens on a plane. Introducing polar coordinates (r, θ) in that plane, with the origin at the center of the spherically symmetric mass M, we see from the expression of the metric in polar coordinates that the square of the velocity of the particle is

$$v^2 = \left(\frac{\Delta s}{\Delta t}\right)^2 = \left(\frac{\Delta r}{\Delta t}\right)^2 + r^2 \left(\frac{\Delta \theta}{\Delta t}\right)^2.$$

So the mechanical energy is given by

$$E = \frac{1}{2}\left(\frac{\Delta r}{\Delta t}\right)^2 + \frac{r^2}{2}\left(\frac{\Delta \theta}{\Delta t}\right)^2 - \frac{M}{r}.$$

Moreover, one can show that the particle's *angular momentum* (per unit mass),

$$L = r^2 \frac{\Delta \theta}{\Delta t},$$

is also conserved along its motion.

We can rewrite the conservation laws as

$$\frac{\Delta \theta}{\Delta t} = \frac{L}{r^2};$$

$$\frac{\Delta r}{\Delta t} = \pm\sqrt{2E + \frac{2M}{r} - \frac{L^2}{r^2}}.$$

These equations are examples of *differential equations*, and can be used to determine all possible motions of m. The idea is as follows: suppose that we are interested in the motion with certain values of E and L which passes through the point with coordinates (r_0, θ_0) at time $t = 0$. We can use the differential equations above to compute approximate values $r_1 = r_0 + \Delta r_0$ and $\theta_1 = \theta_0 + \Delta \theta_0$ for (r, θ) at time $t = \Delta t$, where

Fig. 4.1 Kepler laws

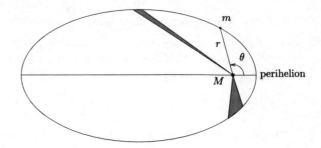

$$\Delta\theta_0 = \frac{L\Delta t}{r_0^2};$$

$$\Delta r_0 = \Delta t \sqrt{2E + \frac{2M}{r_0} - \frac{L^2}{r_0^2}}$$

(we have assumed $\frac{\Delta r}{\Delta t} \geq 0$). Repeating this procedure with r_1 and θ_1, we can compute approximate values $r_2 = r_1 + \Delta r_1$ and $\theta_2 = \theta_1 + \Delta\theta_1$ for (r, θ) at time $t = 2\Delta t$, and, in general, approximate values $r_n = r_{n-1} + \Delta r_{n-1}$ and $\theta_n = \theta_{n-1} + \Delta\theta_{n-1}$ for (r, θ) at time $t = n\Delta t$ for each $n \in \mathbb{N}$. The approximation can be made increasingly better by decreasing Δt.

The differential equations above were solved exactly by Newton,[1] Fig. 4.2 who discovered that the orbits with negative energy are actually *ellipses*, with the center of M at one of the foci (*first Kepler[2] law*). Moreover, the law of conservation of angular momentum has the geometrical interpretation that the line segment joining the center of M to the point particle m sweeps equal areas in equal times (*second Kepler law* – see Exercise 4). In particular, m moves more quickly when it is closer to M (Fig. 4.1). Kepler had deduced his laws empirically from astronomical observations of the planets' motions. Newton showed that these laws were actually mathematical consequences of his universal gravitation law.

4.5 Circular Orbits

From the differential equation for r it is clear that we must have

$$2E + \frac{2M}{r} - \frac{L^2}{r^2} \geq 0 \Leftrightarrow 2Er^2 + 2Mr - L^2 \geq 0.$$

---

[1] Sir Isaac Newton (1643–1727), English physicist, mathematician, astronomer, philosopher and alchemist.

[2] Johannes Kepler (1571–1630), German mathematician, astronomer and astrologer.

Fig. 4.2 Kepler and Newton

If $E < 0$, we see that r can only vary between the two roots of the above polynomial,

$$r_\pm = \frac{-M \pm \sqrt{M^2 + 2EL^2}}{2E},$$

corresponding to the the two points on the ellipse which are closest and furthest to the center of M (both on the major axis). The point which is closest to the center of M is called the orbit's *perihelion*.[3] If $r_- = r_+$, that is, if

$$E = -\frac{M^2}{2L^2},$$

then the orbit must be a circle of radius

$$r = -\frac{M}{2E} = \frac{L^2}{M}.$$

So we have for circular orbits

$$\left(\frac{\Delta\theta}{\Delta t}\right)^2 = \frac{M}{r^3}.$$

In particular, the orbital speed is given by

$$v = r\frac{\Delta\theta}{\Delta t} = \sqrt{\frac{M}{r}}.$$

Note that the escape velocity at a given point is $\sqrt{2}$ times the speed of the circular orbit through that point. We can therefore think of the circular orbit speed

---

[3] Rigorously, this point is called pericenter, and should only be called perihelion for orbits around the Sun; however this abuse of language is frequent.

as a kind of characteristic speed of the gravitational field at that point. Hence one expects relativistic effects to become important only when this speed becomes comparable to the speed of light. For familiar gravitational fields (say the Earth's or the Sun's), this speed is much smaller than the speed of light (so $\frac{M}{r} \ll 1$).

4.6 Important Formulas

- Gravitational acceleration:

$$g = \frac{M}{r^2}$$

- Gravitational potential:

$$\phi = -\frac{M}{r}$$

- Escape velocity:

$$v = \sqrt{\frac{2M}{r}}$$

- Speed of a circular orbit:

$$v = \sqrt{\frac{M}{r}}$$

4.7 Exercises

1. Compute the mass of the Earth in geometrized units from the gravitational acceleration at the Earth's surface (9.8 m/s^2).
2. Compute the escape velocity from the Earth's surface.
3. The orbit of Halley's comet[4] is a very elongated ellipse, which makes its venture away from the Sun almost as far as Pluto's orbit. Therefore Halley's comet (like most comets) moves almost at the Solar System's escape velocity. Knowing that the perihelion of its orbit is about 4.9 light-minutes from the Sun, calculate the Halley comet's speed at that point.
4. Referring to Fig. 4.1, show that when the coordinate θ varies by a small amount $\Delta\theta$, the line segment joining the center of M to the point particle m sweeps an

---

[4] Edmond Halley (1656–1742), English astronomer, geophysicist, mathematician, meteorologist and physicist.

Fig. 4.3 Stars orbiting
Sagittarius A^* (from "Stellar
Orbits Around the Galactic
Center Black Hole", by
A. M. Ghez, S. Salim,
S. D. Hornstein, A. Tanner,
J. R. Lu, M. Morris,
E. E. Becklin and
G. Duchene, The
Astrophysical Journal 620
(2005))

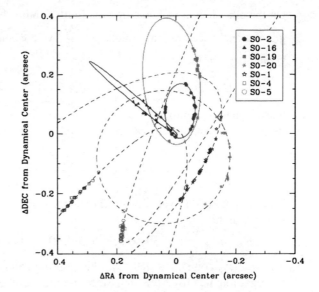

approximate area $\Delta A = \dfrac{1}{2}r^2 \Delta\theta$. Conclude that the law of conservation of
angular momentum can be geometrically interpreted as the statement that the
velocity at which this area is swept is constant.

5. Use the period of the Earth's orbit to compute the mass of the Sun in geometrized units.
6. Calculate the period of a (circular) low Earth orbit.
7. Determine the radius of a geostationary orbit (that is, a circular orbit around the
   Earth with period equal to 24 hr). Do the same for the GPS satellites' orbits,
   whose period is about 12 hr.
8. Compute the period of the Moon's orbit. (Distance from the Earth to the Moon:
   approximately 1.3 light-seconds).
9. Figure 4.3 shows the orbits of a few stars about *Sagittarius* A^*, the black hole at
   the center of our galaxy, which is approximately 26, 000 light-years away. The
   orbits were deduced from the observations plotted in the figure, done during the
   period 1995–2003. Use this data to estimate the mass of Sagittarius A^*.

4.8 Solutions

1. In geometrized units, the gravitational acceleration on the Earth's surface is

$$9.8 \times \frac{1\,\mathrm{m}}{(1\,\mathrm{s})^2} \simeq 9.8 \times \frac{1\,\mathrm{m}}{(3 \times 10^8\,\mathrm{m})^2} \simeq 1.1 \times 10^{-16}\,\mathrm{m}^{-1}.$$

On the other hand, we know that the Earth's radius is $r \simeq 6,400$ km. From the gravitational acceleration formula, $g = \dfrac{M}{r^2}$, we conclude that the Earth's mass in geometrized units is

$$M = gr^2 \simeq 1.1 \times 10^{-16} \times (6.4 \times 10^6)^2 \simeq 4.5 \times 10^{-3}\, \text{m},$$

that is, about 4.5 mm.

2. In geometrized units, the escape velocity from the Earth's surface is

$$v = \sqrt{\frac{2M}{r}} \simeq \sqrt{\frac{9.0 \times 10^{-3}}{6.4 \times 10^6}} \simeq 3.8 \times 10^{-5}.$$

We can obtain this velocity in kilometers per second by multiplying by the speed of light:

$$v \simeq 3.8 \times 10^{-5} \times 3 \times 10^5 \simeq 11\, \text{km/s}.$$

3. The Halley comet's speed will be approximately the escape velocity at 4.9 light-minutes from the Sun. Since the mass of the Sun is about 1.5 km, this speed will then be

$$v = \sqrt{\frac{2M}{r}} \simeq \sqrt{\frac{3.0}{4.9 \times 60 \times 300,000}} \simeq 1.8 \times 10^{-4}.$$

To convert to kilometers per second we multiply by the speed of light:

$$v \simeq 1.8 \times 10^{-4} \times 3 \times 10^5 \simeq 54\, \text{km/s}.$$

4. When the coordinate θ varies by a small amount $\Delta\theta$ in Fig. 4.1, the length r of the line segment joining the center of M to the point particle m practically does not change. So the swept area ΔA is well approximated by the area of the circular sector of radius r and central angle $\Delta\theta$. The area of this sector is a fraction $\dfrac{\Delta\theta}{2\pi}$ of the area of the circle of radius r, that is, $\Delta A = \dfrac{\Delta\theta}{2\pi}\pi r^2 = \dfrac{1}{2}r^2\Delta\theta$. So the orbit's angular momentum is $L = 2\dfrac{\Delta A}{\Delta t}$, and the law of conservation of angular momentum is equivalent to the statement that the velocity at which the area is swept is constant.

5. Since the Earth's orbit is approximately circular, its orbital speed is

$$v = \sqrt{\frac{M}{r}},$$

where M is the Sun's mass and r is the distance from the Earth to the Sun. On the other hand, the period of the Earth's orbit is $T = 1$ year, and the distance from the Earth to the Sun is $r \simeq 8.3$ light-minutes. Using minutes and light-minutes as units, we see that the Earth's orbital speed is

$$v = \frac{2\pi r}{T} \simeq \frac{2\pi \times 8.3}{365 \times 24 \times 60} \simeq 1.0 \times 10^{-4}.$$

Hence the Sun's geometrized mass is

$$M = v^2 r \simeq 10^{-8} \times 8.3 \times 60 \times 3 \times 10^5 \simeq 1.5\,\text{km}.$$

6. For a circular low Earth orbit we have $r \simeq 6,400$ km. The period of the orbit is

$$T = \frac{2\pi r}{v} = 2\pi \sqrt{\frac{r^3}{M}}.$$

Since $\dfrac{M}{r^2} = g \simeq 9.8\,\text{m/s}^2$, we have

$$T \simeq 2\pi \times \sqrt{\frac{6.4 \times 10^6}{9.8}} \simeq 5,100\,\text{s} \simeq 85\,\text{min}.$$

7. If T is the period of a circular orbit of radius r then

$$vT = 2\pi r \Leftrightarrow T\sqrt{\frac{M}{r}} = 2\pi r \Leftrightarrow r = \left(\frac{MT^2}{4\pi^2}\right)^{\frac{1}{3}}.$$

For a geostationary orbit we have $M \simeq 4.5$ mm and $T \simeq 24$ hr. Converting to kilometers yields

$$r \simeq \left(\frac{4.5 \times 10^{-6} \times (24 \times 3,600 \times 300,000)^2}{4\pi^2}\right)^{\frac{1}{3}} \simeq 42,000\,\text{km}.$$

For the GPS satellites' orbits $T \simeq 12$ hr, and hence

$$r \simeq \left(\frac{4.5 \times 10^{-6} \times (12 \times 3,600 \times 300,000)^2}{4\pi^2}\right)^{\frac{1}{3}} \simeq 27,000\,\text{km}.$$

8. For the Moon's (circular) orbit we have $r \simeq 1.3$ light-seconds and $M \simeq 4.5$ mm. Converting to seconds, the period of the orbit is

$$T = \frac{2\pi r}{v} = 2\pi \sqrt{\frac{r^3}{M}} \simeq 2\pi \sqrt{\frac{1.3^3 \times 3 \times 10^5}{4.5 \times 10^{-6}}} \simeq 2.4 \times 10^6\,\text{s},$$

that is, about 28 days.

9. From Fig. 4.3 we see that the orbit of the star SO–20 is roughly circular, with radius approximately 0.2 arcsec wide, corresponding to a distance

$$\frac{0.2}{3,600} \times \frac{\pi}{180} \times 26,000 \simeq 2.52 \times 10^{-2}\,\text{light-years}.$$

This distance is about

$$\frac{2.52 \times 10^{-2} \times 365 \times 24 \times 60}{8.3} \simeq 1,600$$

times the distance from the Earth to the Sun. If T is the period of a circular orbit of radius r then

$$vT = 2\pi r \Leftrightarrow T\sqrt{\frac{M}{r}} = 2\pi r \Leftrightarrow M = \frac{4\pi^2 r^3}{T^2}.$$

Since the star SO–20 took eight years to complete about a quarter of a revolution, its period is around 32 years. We conclude that the mass of Sagittarius A^* is approximately

$$\frac{1,600^3}{32^2} \simeq 4 \times 10^6$$

times the mass of the Sun.

Chapter 5
General Relativity

In this chapter we introduce Einstein's general theory of relativity. The key idea is the equivalence principle, according to which a (small) free-falling frame is equivalent to an inertial frame. This principle can be used to make predictions about the behavior of light in a gravitational field; in particular, we show that the period of a light signal increases as it moves upwards (gravitational redshift). We argue that this fact is fundamentally incompatible with Minkowski geometry, and can only be accommodated by accepting that space–time is curved. The equivalence principle then becomes the observation that curved space–time is locally flat, and implies that free-falling particles must move along geodesics (and light rays along null geodesics) just like in flat Minkowski space–time. Given the matter distribution, the space–time metric can be found by solving the Einstein equation, whose nature we describe.

5.1 Equivalence Principle

In 1907, just two years after publishing the special theory of relativity, Einstein had, in his own words, "the happiest thought of his life". This idea, which he later called the *equivalence principle*, was simply this—for a free-falling observer everything happens as if there was no gravity at all.

To understand what Einstein meant we have to remember that the gravitational acceleration is the *same* for all bodies, regardless of their mass. This is illustrated by Galileo's legendary experiment of dropping balls of different weights from the leaning tower of Pisa and watching them reach the ground simultaneously. Should Galileo have jumped together with the balls he would have seen them floating around him, and could momentarily imagine that he was in a gravity-free environment. This is exactly what happens to astronauts in orbit (Fig. 5.1). It is sometimes said that there is no gravity in orbit, but this is obviously wrong—if there was no gravity there would be no forces acting on the spacecraft, and it

J. Natário, *General Relativity Without Calculus*, Undergraduate Lecture
Notes in Physics, DOI: 10.1007/978-3-642-21452-3_5,
© Springer-Verlag Berlin Heidelberg 2011

Fig. 5.1 An orbiting
astronaut. (Image credit:
NASA)

Fig. 5.1 An orbiting astronaut. (Image credit: NASA)

would just move away from Earth on a straight line with constant velocity according to the law of inertia. Actually, an orbiting spacecraft is free-falling around the Earth; that is why its crew seem to float inside.

5.2 Gravitational Redshift

The first thought experiment to which Einstein applied his equivalence principle was the following—suppose that Einstein, from an intermediate floor on the leaning tower of Pisa, sends a light signal with period T towards Galileo, who is on the top floor. What is the period T' measured by Galileo?

To answer this question let us suppose that both Einstein and Galileo jump from the tower as the signal is emitted. Then they would be in a free-falling frame, thus equivalent to an inertial frame (without gravity). So Galileo would measure the same period T for the light signal (as he would be at rest with respect to Einstein in this inertial frame). Let Δz be the vertical distance from Einstein to Galileo, and g the gravitational acceleration. It takes the light signal a time Δz to get to the top floor (as $c = 1$), and in this time interval Galileo acquires a speed $v = g\Delta z$ with respect to the tower (we are assuming that this speed is much smaller than the speed of light, $v \ll 1$, so that this Newtonian formula is approximately correct; we are also assuming that the light signal's period is much smaller than the time taken by the signal to reach the top floor, $T \ll \Delta z$). According to the Doppler effect formula for $|v| \ll 1$ (see Exercise 5 in Chap. 2), we must have approximately

$$T' = (1 + g\Delta z)T,$$

that is, should Galileo have remained on the top floor he would have measured a *bigger* period for the light signal. Since the period of visible light increases from blue to red, it is usual to call this effect the *gravitational redshift*.

Fig. 5.2 Minkowski
geometry is incompatible
with the gravitational redshift

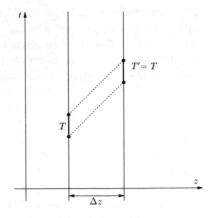

Recall that the potential energy of a particle on the Earth's surface is $U = mgz$, and so the corresponding gravitational potential is $\phi = gz$. Another way to write the gravitational redshift formula is then

$$T' = (1 + \Delta\phi)T.$$

As long as $|\Delta\phi| \ll 1$, this formula is still valid even when the field is not uniform (one just has to introduce sufficiently many intermediate observers, for whom the field is approximately uniform). Since the two observers are at rest with respect to each other, this formula can only be interpreted as meaning that time flows at different rates for observers in different points of a gravitational field—observers in lower positions measure smaller time intervals.

5.3   Curved Space–Time

Figure 5.2 clearly shows that the gravitational redshift is not compatible with Minkowski geometry, according to which any two observers at rest on a given inertial frame must measure the same period for any light signal. To account for the gravitational redshift we need a space–time diagram which distorts intervals, just like any map of the sphere distorts distances. This led Einstein to conjecture that the existence of a gravitational field would be equivalent to space–time being curved.

A small region of a curved surface (say a sphere) looks approximately flat,[1] the approximation becoming better as the region becomes smaller—for example, the sum of a triangle's internal angles tends to π as the triangle becomes smaller and smaller, since the difference is given by the curvature multiplied by the triangle's area. In the same way, thought Einstein, a small region of a curved space–time

---

[1] It is partly for this reason that for some time it was believed that the Earth is flat—the portion of the Earth's surface we can see is small when compared with the radius of the Earth.

must look approximately as flat Minkowski space–time. But this is exactly the equivalence principle—in a small region of a free-falling frame everything happens as if we were in an inertial frame (the region must be small so that the gravitational field is approximately uniform across that region; otherwise not all bodies would fall with the same acceleration and it would be possible to infer the existence of a gravitational field). Moreover, any free-falling body in one of these free-falling frames will move on a straight line with constant velocity, and hence, as seen in Chap. 2, its history will be a maximum length causal curve. We conclude that in a curved space–time the histories of free-falling bodies must be maximum length causal curves, that is, *geodesics*. In the same way, in a small free-falling frame the history of a light ray must be a causal curve of zero length with the property that it is the *unique* causal curve joining any two of its points. We conclude that in a curved space–time the histories of light rays are causal curves of this kind, called *null geodesics*.

These ideas were already clear for Einstein in 1912. To complete his theory of *general relativity* he now needed to know how to compute the space–time metric determined by a given matter distribution. For that, Einstein spent the next three years studying the mathematics of curved spaces, with the help of his mathematician friend Marcel Grossmann[2] Fig. 5.3. There was no method for (nor any guarantee of) reaching the right result—he literally needed to *guess* the answer. After many blind alleys and failed attempts, he finally managed to find the correct equation, by the end of 1915 (Einstein was almost scooped by the famous mathematician Hilbert,[3] who arrived at the same equation by a different method almost at the same time). The *Einstein equation* Fig. 5.4 is

$$\mathbf{R} = 8\pi\mathbf{T}$$

where **R** and **T** are mathematical objects called *tensors* (a generalization of vectors). The tensor **R** is called the *Ricci curvature tensor*,[4] and describes (part of) the space–time curvature at each point.[5] It can be computed from the variation of the space–time metric coefficients in each direction. The tensor **T** is called the *trace-reversed energy–momentum tensor*, and describes the matter distribution. Thus the Einstein equation specifies how matter curves space–time. It turns out that **R** and **T** have ten independent components, and so the Einstein equation is in fact a complicated set of ten differential equations for the space–time metric coefficients.

---

[2] Marcel Grossmann (1878–1936), Hungarian mathematician.

[3] David Hilbert (1862–1943), German mathematician.

[4] Gregorio Ricci-Curbastro (1853–1925), Italian mathematician.

[5] For surfaces this tensor has just one independent component, which is twice the curvature, but it is more complicated for higher dimensional spaces.

Fig. 5.3 Grossmann, Hilbert and Ricci

Fig. 5.4 Einstein writing his equation for vacuum ($\mathbf{R} = \mathbf{0}$)

5.4 Important Formulas

- Gravitational redshift:

$$\boxed{T' = (1 + \Delta\phi)T \qquad \text{for} \quad |\Delta\phi| \ll 1}$$

- Planck–Einstein relation (see Exercise 2):

$$\boxed{E = \frac{h}{T}}$$

- Einstein's mass–energy equivalence relation (see Exercise 2):

$$\boxed{E = mc^2}$$

5.5 Exercises

1. Since in everyday situations the gravitational redshift is very small, this effect
   was only experimentally confirmed in the beginning of the sixties. The first
   experiment was carried out in an elevator shaft of one of the Harvard university
   buildings, which was 23 m high. Compute the percentual variation of the period
   measured in this experiment.
2. Sometimes light behaves as if composed of particles, called *photons*, with
   energy given by the *Planck[6]-Einstein relation*

$$E = \frac{h}{T},$$

   where h is *Planck's constant* and T is the light's period. From the *mass–energy
   equivalence relation*

$$E = mc^2$$

   (which Einstein derived as a consequence of special relativity), one would
   expect a photon going up in a gravitational field to lose an amount of energy

$$\Delta E = \frac{E}{c^2} \Delta \phi.$$

   Show that if this is the case then the Planck–Einstein relation implies the
   gravitational redshift formula.
3. Correct the result of Exercise 13 in Chap. 1 to include the effect of the gravitational field. Assume that the airplane flew at an average altitude of 10 km.
4. The global positioning system (GPS) uses satellites on 12-h orbits carrying very
   precise atomic clocks. It is very important that these clocks are synchronized
   with the clocks on the ground tracking stations, since any desynchronization
   will result in positional errors of the same magnitude (using $c = 1$). Show that
   if one failed to apply relativistic corrections then the desynchronization with
   respect to a ground station on the equator would be about 12 light-kilometers
   after just one day.
5. Due to its rotation motion the Earth is not a perfect sphere, being flattened at the
   poles. Because of this the Earth's gravitational potential is given by a more
   complicated expression than simply $\phi = -\frac{M}{r}$ (which is however a very good
   approximation). Actually, $\phi - \frac{v^2}{2}$ has the same value at all points on the surface
   of the Earth, where ϕ is the Earth's gravitational potential and v is the Earth's
   rotation speed at that point. Show that as a consequence of this formula all
   clocks on the surface of the Earth tick at the same rate.

---

[6] Max Planck (1858–1947), German physicist.

6. *Twin paradox (yet again)*: Recall once more the setup of Exercise 15 in Chap. 1. Two twins, Alice and Bob, part on their 20th birthday. While Alice remains on Earth (which is an inertial frame to a very good approximation), Bob departs at 80% of the speed of light towards Planet X, 8 light-years away from Earth. Therefore Bob reaches his destination 10 years later (as measured on the Earth's frame). After a short stay, he returns to Earth, again at 80% of the speed of light. Consequently Alice is 40 years old when she sees Bob again, whereas Bob is only 32 years old.

   (a) In both legs of his journey Bob is on inertial frames, and the time dilation formula applies. How much time does he expect Alice to experience?

   (b) Bob is then forced to conclude that Alice experienced the missing time during the very short (according to him) acceleration phase of his journey. Check that this is consistent with the gravitational redshift formula.

5.6 Solutions

1. The percentual variation was

$$\frac{T' - T}{T} = \Delta\phi = g\Delta z \simeq \frac{9.8 \times 23}{(3 \times 10^8)^2} \simeq 2.5 \times 10^{-15}.$$

2. Let E and T be the energy and the period of a photon on a given point P of the gravitational field, and let E', T' be the same quantities at another point P'. Let $\Delta\phi$ be the potential difference between P' and P. The Planck–Einstein relation implies

$$ET = E'T',$$

and if we accept the formula for the photon's energy loss we are force to conclude that

$$E' = E - \frac{E}{c^2}\Delta\phi = \left(1 - \frac{\Delta\phi}{c^2}\right)E.$$

Therefore

$$T = \left(1 - \frac{\Delta\phi}{c^2}\right)T',$$

or, using the fact that $\left|\dfrac{\Delta\phi}{c^2}\right| \ll 1$, we have the approximate formula

$$T' = \left(1 + \frac{\Delta\phi}{c^2}\right)T,$$

which is just the gravitational redshift formula ($c = 1$ in our units).

3. Since the airplane flew at an average altitude of 10 km, the clocks on board the airplane registered a longer travel time than similar clocks on the surface by a fraction

$$\Delta\phi = g\Delta z \simeq \frac{9.8 \times 10^4}{(3 \times 10^8)^2} \simeq 1.1 \times 10^{-12}.$$

Since the travel took about 124,000 s, this corresponds to an additional time of about

$$1.1 \times 10^{-12} \times 1.24 \times 10^5 \simeq 1.4 \times 10^{-7} \text{s},$$

that is, about 140 ns (independently of the flight direction). Therefore the clock which flew eastwards was only late about $170 - 140 = 30$ ns, whereas the clock which flew westwards was about $80 + 140 = 220$ ns early.[7]

4. If an inertial observer "at infinity" (i.e. far away from the Earth) measures a time interval Δt, a satellite moving with speed v on a point at a distance r from the center of the Earth measures a time interval

$$\Delta t_{SAT} = \sqrt{1 - v^2}\left(1 - \frac{M}{r}\right)\Delta t \simeq \left(1 - \frac{v^2}{2}\right)\left(1 - \frac{M}{r}\right)\Delta t \simeq \left(1 - \frac{v^2}{2} - \frac{M}{r}\right)\Delta t.$$

In the same way, an observer on the Earth's surface measures a time interval

$$\Delta t_{EARTH} \simeq \left(1 - \frac{V^2}{2} - \frac{M}{R}\right)\Delta t,$$

where V is the Earth's rotation speed and R is the radius of the Earth. Therefore

$$\frac{\Delta t_{SAT}}{\Delta t_{EARTH}} = \frac{1 - v^2/2 - M/r}{1 - V^2/2 - M/R} \simeq \left(1 - \frac{v^2}{2} - \frac{M}{r}\right)\left(1 + \frac{V^2}{2} + \frac{M}{R}\right)$$

$$\simeq 1 - \frac{v^2}{2} - \frac{M}{r} + \frac{V^2}{2} + \frac{M}{R}.$$

We have already seen that $r \simeq 27,000$ km, so that

$$\frac{M}{r} \simeq \frac{4.5 \times 10^{-6}}{27,000} \simeq 1.7 \times 10^{-10}.$$

On the other hand,

$$\frac{v^2}{2} = \frac{M}{2r} \simeq 0.8 \times 10^{-10}.$$

---

[7] The actual predicted values for the difference with respect to the stationary clocks were -40 ± 23 and 275 ± 21 ns, whereas the measured values were -59 ± 10 and 273 ± 7 ns. The difference with respect to our values is due to our simplifying assumptions (constant speed, latitude and altitude).

Analogously, since $R \simeq 6,400$ km, we have

$$\frac{M}{R} \simeq \frac{4.5 \times 10^{-6}}{6,400} \simeq 7.0 \times 10^{-10}.$$

Finally, the Earth's rotation speed on the equator is about

$$\frac{2\pi \times 6,400}{24 \times 3,600} \simeq 0.47 \text{ km/s},$$

so that

$$\frac{V^2}{2} \simeq 1.2 \times 10^{-12}.$$

We conclude that

$$\frac{\Delta t_{\text{SAT}}}{\Delta t_{\text{EARTH}}} \simeq 1 + 4.5 \times 10^{-10},$$

that is, the clock on the satellite is fast by about $4.5 \times 10^{-10} \times 24 \times 3,600 \simeq 4.0 \times 10^{-5}$ s per day, corresponding to about $4.0 \times 10^{-5} \times 3 \times 10^5 = 12$ light-kilometers.

5. If an inertial observer "at infinity" (i.e. far away from the Earth) measures a time interval Δt, an observer moving with speed v on a point where the Earth's gravitational potential is ϕ measures a time interval

$$\Delta t' = \sqrt{1 - v^2}(1 + \phi)\Delta t \simeq \left(1 - \frac{v^2}{2}\right)(1 + \phi)\Delta t \simeq \left(1 - \frac{v^2}{2} + \phi\right)\Delta t.$$

Since $\phi - \frac{v^2}{2}$ is equal for all points on the surface of the Earth, we conclude that all clocks on the surface of the Earth tick at the same rate. For instance, a clock on the equator is moving faster than a clock on the north pole (so it should tick slower), but it is further away from the center of the Earth (so it should tick faster); these two effects exactly cancel.

6. (a) We have already seen that Bob takes 6 years to complete the first leg of the journey. Since during this time Alice is moving at 80% of the speed of light with respect to him, he expects her to experience

$$6\sqrt{1 - 0.8^2} = 6 \times 0.6 = 3.6 \text{ years}.$$

The same is true for the return leg, and so he expects Alice to experience in total $3.6 + 3.6 = 7.2$ years.

   (b) Let a be Bob's acceleration and Δt the (small) time interval he spends accelerating. Then we must have $a\Delta t = 0.8 + 0.8 = 1.6$. When Bob is accelerating he can imagine that he is in an uniform gravitational field, with Alice 8 light-years higher up, corresponding to a potential difference $\Delta\phi = 8a$. By the gravitational redshift formula, she must experience a time interval

$$\Delta t' = (1 + \Delta\phi)\Delta t = \Delta t + 8a\Delta t = \Delta t + 8 \times 1.6 \simeq 12.8 \text{ years.}$$

Notice that this is exactly the missing time, $12.8 + 7.2 = 20$. (The fact that this works out perfectly is however just a happy coincidence—the gravitational redshift formula can only be expected to hold for $|\Delta\phi| \ll 1$. Moreover, Alice is not at rest in the accelerated frame—she must be in free fall as she is an inertial observer).

Chapter 6
The Schwarzschild Solution

In this chapter we study the Schwarzschild metric, corresponding to the gravitational field of a spherically symmetric body of mass M. Unlike Minkowski spacetime, it contains a preferred class of observers, called stationary observers, who are not moving with respect to the central mass. We compute their proper time in terms of the time coordinate t and use it to obtain the exact redshift formula ("gravity delays time"). We also explain how by measuring distances between them stationary observers are led to the conclusion that space is curved ("gravity curves space"). Next we write the differential equations for the geodesics and see how they differ subtly from the Newtonian differential equations for free-falling motion. This leads to orbits that are approximately ellipses but whose axes slowly rotate. This effect, which had been observed for the planet Mercury, was the first triumph of the general theory of relativity. The differential equations for the null geodesics (light rays) also predict two new effects: a bending of the light rays passing near the spherically symmetric body (whose experimental confirmation in 1919 led to a wide acceptance of the theory), and a delay in the time of arrival of the light, known as the Shapiro delay (experimentally confirmed in 1966). Finally, we analyze the surface $r = 2M$, where the Schwarzschild metric is not defined. We show that this is a problem with the choice of coordinates, and not the space-time itself, and introduce the so-called Painlevé coordinates, which remove this problem. The surface $r = 2M$, however, remains special: it marks the boundary of a region from which nothing, not even light, can escape–a black hole.

J. Natário, *General Relativity Without Calculus*, Undergraduate Lecture
Notes in Physics, DOI: 10.1007/978-3-642-21452-3_6,
© Springer-Verlag Berlin Heidelberg 2011

Fig. 6.1 Karl Schwarzschild

6.1 The Schwarzschild Solution

In 1916, just a few months after the publication of the Einstein equation, Schwarzschild[1] Fig. 6.1, who was then on the Russian front, discovered the solution corresponding to the gravitational field of a spherically symmetric body of mass M. If we restrict ourselves to events on the equatorial plane, the Schwarzschild metric is

$$\Delta\tau^2 = \left(1 - \frac{2M}{r}\right)\Delta t^2 - \left(1 - \frac{2M}{r}\right)^{-1}\Delta r^2 - r^2\Delta\theta^2$$

(for $r > 2M$).

How are we to interpret the coordinates (t, r, θ)? We notice that when $M = 0$ the Schwarzschild metric becomes

$$\Delta\tau^2 = \Delta t^2 - \Delta r^2 - r^2\Delta\theta^2,$$

which is just the Minkowski metric in polar coordinates, since, as we have seen, the metric of the Euclidean plane is

$$\Delta s^2 = \Delta x^2 + \Delta y^2 = \Delta r^2 + r^2\Delta\theta^2.$$

This makes sense: when $M = 0$ there is no gravity and therefore space-time must be flat Minkowski space-time. We can then think of the Schwarzschild coordinates (t, r, θ) as a generalization of the Minkowski coordinates (t, r, θ).

---

[1] Karl Schwarzschild (1873–1916), German physicist and astronomer.

6.2 Stationary Observers

The curves along which the coordinates r and θ are constant are the histories of observers at rest with respect to the spherically symmetric body M, which we shall call *stationary observers*. Note that these curves *are not* geodesics, since these observers are not in free fall (if they were, their coordinate r should decrease). Actually, it can be shown that these observers measure a gravitational acceleration

$$g = \frac{\dfrac{M}{r^2}}{\sqrt{1 - \dfrac{2M}{r}}},$$

which is approximately the Newtonian result for $\frac{M}{r} \ll 1$.

When the coordinate t varies by Δt, a stationary observer measures a proper time interval

$$\Delta\tau = \Delta t\sqrt{1 - \frac{2M}{r}}$$

(since $\Delta r = \Delta\theta = 0$). Note that when r is very big we have $\Delta\tau \simeq \Delta t$. Therefore we can interpret the coordinate t as the time measured by a stationary observer far away from M ("at infinity").

6.3 Redshift

The Schwarzschild metric is invariant under time translations. This means that the interval between two nearby events P_1 and P_2, with coordinates (t_1, r_1, θ_1) and (t_2, r_2, θ_2). is equal to the interval between the events Q_1 and Q_2 with coordinates $(t_1 + \Omega, r_1, \theta_1)$ and $(t_2 + \Omega, r_2, \theta_2)$, for any $\Omega \in \mathbb{R}$. So if γ is a geodesic, that is, a curve of maximum length, the curve γ' obtained from γ by moving all of its points to points with the coordinate t increased by Ω is also a geodesic. The same is true if γ is a null geodesic.

Consider two stationary observers O and O', with radial coordinates r and r'. Suppose that O sends a light signal with period T towards O'. The history of the light ray corresponding to the beginning of the period is a null geodesic γ. The light ray corresponding to the end of the period is another null geodesic γ', constructed from γ translating by Ω in the t coordinate (Fig. 6.2). The period T measured by O is related to Ω by

$$T = \Omega\sqrt{1 - \frac{2M}{r}}.$$

In the same way, the period T' measured by O' satisfies

Fig. 6.2 Redshift in the
Schwarzschild geometry

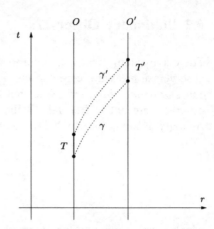

$$T' = \Omega\sqrt{1 - \frac{2M}{r'}}.$$

We conclude that

$$T' = T\sqrt{\frac{1 - \dfrac{2M}{r'}}{1 - \dfrac{2M}{r}}}.$$

This is the redshift formula for stationary observers in the Schwarzschild metric.
It reduces to the approximate formula in Chap. 5 for $\frac{M}{r}, \frac{M}{r'} \ll 1$ (see Exercise 1).
Note that if $r' > r$ then $T' > T$, as one would expect. This statement is sometimes
summarized by saying that *gravity delays time*.

6.4 Space Curvature

Suppose that two nearby stationary observers, with spatial coordinates (r, θ) and
$(r + \Delta r, \theta + \Delta\theta)$, want to measure the distance between them. To do so they just
have to measure the time taken by a light signal to travel from one to the other
($c = 1$). A light signal propagating between the two observers takes a coordinate
time interval Δt such that

$$\left(1 - \frac{2M}{r}\right)\Delta t^2 - \left(1 - \frac{2M}{r}\right)^{-1}\Delta r^2 - r^2\Delta\theta^2 = 0.$$

However, the stationary observers measure a proper time interval

$$\Delta\tau = \Delta t\sqrt{1 - \frac{2M}{r}},$$

Fig. 6.3 Surface with the metric measured by the stationary observers

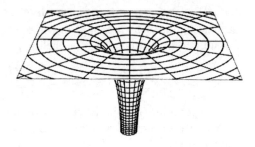

and so measure a distance

$$\Delta s = \Delta \tau = \Delta t \sqrt{1 - \frac{2M}{r}} = \sqrt{\left(1 - \frac{2M}{r}\right)^{-1} \Delta r^2 + r^2 \Delta \theta^2},$$

that is,

$$\Delta s^2 = \left(1 - \frac{2M}{r}\right)^{-1} \Delta r^2 + r^2 \Delta \theta^2.$$

This is not the metric of an Euclidean plane, since the coefficient of Δr^2 is not 1 (although it does approach 1 for large r). In fact, it can be shown to be the metric of the surface in Fig. 6.3. In other words, stationary observers on the equatorial plane deduce from their distance measurements that they are actually living on a curved surface. This statement is sometimes summarized by saying that *gravity curves space*.

6.5 Orbits

To compute the histories of free-falling particles we have to compute the geodesics of the Schwarzschild metric. One can show that these satisfy the following differential equations:

$$\frac{\Delta \theta}{\Delta \tau} = \frac{L}{r^2};$$

$$\frac{\Delta r}{\Delta \tau} = \pm \sqrt{2E + \frac{2M}{r} - \left(1 - \frac{2M}{r}\right) \frac{L^2}{r^2}};$$

$$\frac{\Delta t}{\Delta \tau} = \left(1 - \frac{2M}{r}\right)^{-1} \sqrt{1 + 2E}.$$

Note that the first two equations are almost identical to the Newtonian equations, with the proper time τ playing the role of the Newtonian time. This means that if the relativistic effects are small ($\frac{M}{r} \ll 1$) then the trajectories of free-falling particles in the Schwarzschild metric are approximately ellipses. However, they are not exactly ellipses: their perihelion *precesses* (i.e. rotates), at a rate of about

$$\frac{6\pi M}{r}$$

radians per orbit (for almost circular orbits of radius r).

Actually, the perihelia of the orbits of all planets in the Solar System precess for other reasons. This is not surprising, since besides the Sun's gravitational force, which is by far the main influence, the planets also feel the gravitational attraction of each other. When Einstein discovered the general theory of relativity, Newton's theory could explain the precession of the perihelia of all planets except one: Mercury. The precession of Mercury's perihelion is about 5,600 arcsec per century; Newton's theory predicted 5,557. There was thus an unexplained discrepancy of 43 arcsec per century.[2] When Einstein computed the precession of Mercury's perihelion due to general relativistic effects he exactly obtained, to his great satisfaction, the missing 43 arcsec per century (see Exercise 2).

By similar methods to those used in the Newtonian theory, one can show that there are circular orbits of any radius $r > 3M$. By coincidence, the Newtonian result

$$\left(\frac{\Delta\theta}{\Delta t}\right)^2 = \frac{M}{r^3}.$$

still holds.

6.6 Light Rays

The histories of light rays correspond to the null geodesics of the Schwarzschild metric, which satisfy the following differential equations:

$$\frac{\Delta\theta}{\Delta\lambda} = \frac{L}{r^2};$$

$$\frac{\Delta r}{\Delta\lambda} = \pm\sqrt{2E - \left(1 - \frac{2M}{r}\right)\frac{L^2}{r^2}};$$

$$\frac{\Delta t}{\Delta\lambda} = \left(1 - \frac{2M}{r}\right)^{-1}\sqrt{2E}.$$

The parameter λ plays the same role as the proper time on a geodesic (recall that null geodesics have zero length). These equations can be obtained from the geodesic equations in the limit $E \gg 1$. This is what one would expect, as particles

---

[2] One of the possible explanations put forward at the time was the the existence of a small planet, called Vulcan, between Mercury and the Sun.

Fig. 6.4 Sir Arthur
Eddington

with very high energies move very close to the speed of light. It is however clear
that these equations *are not* the equations for a null geodesic in Minkowski's
space-time, which can be obtained by setting $M = 0$.

The fact that $M > 0$ makes $\frac{\Delta t}{\Delta \lambda}$ *bigger* than what it would be in Minkowski's
space-time. Therefore a light ray takes a longer coordinate time t to travel a given
path in the Schwarzschild metric. This is the so-called *Shapiro*[3] *effect*, and has
been measured by radar experiments in the Solar System starting in 1966.

On the other hand, $M > 0$ makes the absolute value of $\frac{\Delta r}{\Delta \lambda}$, and hence $\frac{\Delta r}{\Delta \theta}$, *bigger*
than it would be in Minkowski's space-time. Consequently the light ray follows a
curved trajectory, instead of the straight line it would travel for $M = 0$. This is the
so-called *gravitational lens effect*, and was (together with the gravitational redshift
and the precession of Mercury's perihelion) one of the three experimental tests of
general relativity proposed by Einstein. The experimental confirmation of this
prediction, achieved by an English expedition led by Eddington[4] Fig. 6.4 in 1919,
made Einstein a celebrity overnight.

The English astronomers had to travel to the remote locations of the Prince's
Island (then a Portuguese colony) and Sobral (Brazil) to photograph a total eclipse
of the Sun. The reason for this is illustrated in Fig. 6.5: the presence of a body with a
large mass M bends the light rays, shifting the images of objects with respect to their
usual positions. By comparing a picture of the sky around a body of high mass with
a picture of the same region at a time when the body is not there it is then possible to
measure the shift in the stars' positions due to the bending of the light rays.

The bending angle computed by Einstein was

---

[3] Irwin Shapiro (1929–), American astrophysicist.

[4] Sir Arthur Eddington (1882–1944), English astrophysicist.

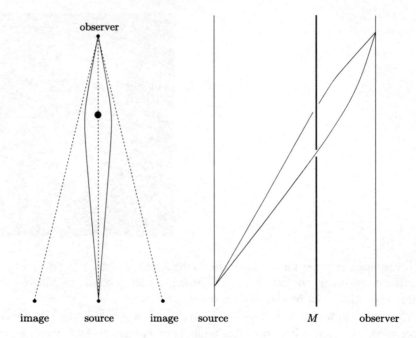

Fig. 6.5 Gravitational lens effect in space and in space-time

$$\frac{4M}{b},$$

where b is the distance of closest approach of the light ray to the body (for comparison's sake, Newtonian gravity predicts *half* this value for a particle moving at the speed of light). This is a very small angle under normal circumstances. For a light ray just grazing the surface of the Sun (which is by far the body of larger mass in the Solar System) it is only about 1.75 arcsec (see Exercise 7). Hence the need of using the Sun to measure this effect, and consequently of photographing the Sun during a total eclipse (to make the stars visible).

Figure 6.5 illustrates the case in which the mass M is directly between the light source and the observer; in this case she sees *two* images, on opposite sides of M. In the space-time diagram this corresponds to the existence of two null geodesics connecting the same events, which, as we saw in Chap. 3, signals the presence of curvature.

A spectacular example of this effect is the so-called *Einstein Cross* (Fig. 6.6). It consists of *four* images of the *same quasar*,[5] 8 billion[6] light-years away, surrounding the nucleus of a galaxy "only" half a billion light-years away, which

---

[5] A *quasar* is an active galaxy, which can be seen across very large (cosmological) distances.

[6] We adopt the standard convention that one billion is a thousand millions (10^9).

Fig. 6.6 Einstein Cross
(image by ESA's Faint
Object Camera on board
NASA's Hubble Space
Telescope)

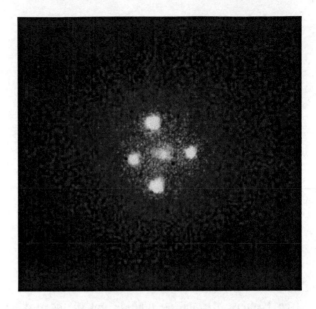

is acting as a gravitational lens.[7] The reason why the images are known to be of the
same quasar is that they show the same exact pattern of brightness variation. Due
to the Shapiro effect, however, the brightness variation of the different images is
desynchronized by a few days.

6.7 Black Holes

The Schwarzschild metric is not defined for $r = 2M$ (the so-called *Schwarzschild
radius*), since there is a division by zero in the coefficient of Δr^2. For a long time it
was not clear what to make of this.

   In the Newtonian theory, the Schwarzschild radius corresponds to the points
where the escape velocity is equal to the speed of light. In the Schwarzschild
metric a similar phenomenon occurs: the escape velocity for a stationary observer
approaches the speed of light as r approaches $2M$ (see Exercise 5). Moreover, the
redshift and the gravitational field measured by these observers approach infinity
as r approaches $2M$. For these reasons it was initially thought that $r = 2M$ was a
mathematical singularity, beyond which one could not extend the Schwarzschild
space-time. However, it was noticed that the curvature of this space-time is

$$\frac{M}{r^3},$$

---

[7] If the alignment was even better one would see *infinite* images, forming a so-called *Einstein
ring*.

Fig. 6.7 Space-time diagram
for the Schwarzschild
solution containing the
histories of: **a** a stationary
observer; **b** a particle falling
through the event horizon;
c the event horizon; **d** the
singularity

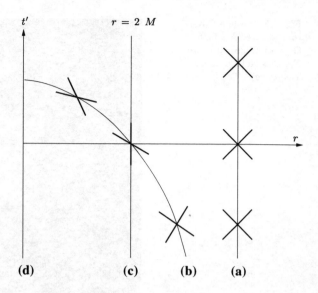

(see Exercise 4), and so nothing out of the ordinary seems to happen at the Schwarzschild radius.

An analogous situation occurs on the map of the sphere obtained from the cylindrical projection: although the coefficient of Δy^2 is not defined for $y = -R$ (south pole), nothing special happens to the curvature of the sphere as y approaches this value, since it is constant equal to $\frac{1}{R^2}$. In this case, it is easy to see that the sphere does not have any singularity at the south pole: the coordinates (x,y) associated to the cylindrical projection just happen to be ill defined at this point. The problem can be overcome by choosing a different coordinate system (for instance the coordinates (x,y) associated to the stereographic projection).

A coordinate system which is well defined on the region $r = 2M$ was discovered by Painlevé[8] in 1921. Painlevé substituted the Schwarzschild coordinate t by a different time coordinate[9] t', in terms of which the metric is written (see Exercise 10)

$$\Delta\tau^2 = \Delta t'^2 - \left(\Delta r + \sqrt{\frac{2M}{r}}\Delta t'\right)^2 - r^2\Delta\theta^2$$

$$= \left(1 - \frac{2M}{r}\right)\Delta t'^2 - 2\sqrt{\frac{2M}{r}}\Delta t'\Delta r - \Delta r^2 - r^2\Delta\theta^2.$$

---

[8]   Paul Painlevé (1863–1933), French mathematician.

[9]   This coordinate is the proper time measured by the family of observers who fall in radially from infinity, duly synchronized (see Exercise 11).

Note that in these coordinates there is no problem at $r = 2M$. However, the coefficient of $\Delta t'^2$ vanishes at the Schwarzschild radius. This means that the curves on which $r = 2M$ and θ is constant are *null geodesics*. No wonder that there cannot be stationary observers at $r = 2M$ (they would have to be moving at the speed of light). Similarly, there cannot exist stationary observers for $r < 2M$, since these would correspond to non-causal curves (they would have to be moving faster than the speed of light).

These facts can be geometrically understood by representing the light cones in a space-time diagram (Fig. 6.7). These are given at each point of coordinates (t', r) by

$$\Delta \tau^2 = 0 \Leftrightarrow \pm \Delta t' = \Delta r + \sqrt{\frac{2M}{r}} \Delta t' \Leftrightarrow \frac{\Delta r}{\Delta t'} = \pm 1 - \sqrt{\frac{2M}{r}}.$$

For $r \gg 2M$ the light cones approach the Minkowski light cones, with slopes ± 1, corresponding to light rays receding from or approaching $r = 0$. As r diminishes, however, the light cones tilt towards $r = 0$. At the Schwarzschild radius, the light cones are so tilted that the light ray "receding" from $r = 0$ actually stands still. For $r < 2M$ the situation is even more dramatic: the light ray "receding" from $r = 0$ is actually approaching $r = 0$. Therefore a particle entering the region $r < 2M$ can never leave; moreover, it is forced to move towards $r = 0$.

Also, it is clear that observers at infinity can never see any event in the region $r \leq 2M$ (see Exercise 12); this region is then called a *black hole*. The surface $r = 2M$, which bounds this region, is called the *event horizon* (since one cannot see beyond it). Finally, the curve $r = 0$ is called the *singularity*. It is indeed a mathematical singularity, beyond which it is not possible to continue the Schwarzschild space-time, since the curvature becomes infinite there.

Physically, the curvature of a space-time measures the so-called *tidal force*, resulting from the non-uniformity of the gravitational field.[10] Consider, for instance, an astronaut in orbit, floating upside-down. Then the Earth's gravitational force on her head is slightly larger than on her feet. Hence there is a residual stretching force (the tidal force). At the singularity inside the black hole this force becomes infinite, and thus any object which falls inside a black hole ends up destroyed.

Ordinary spherically symmetric bodies (stars, planets, golf balls) have radii which are much larger than their Schwarzschild radius. However, it is known that stars which are three times more massive than the Sun at the end of their evolutionary process end up forming black holes. A famous example is the *Cygnus X-1* black hole, nine times more massive than the Sun, which is about 6,000 light-years away.

Besides black holes with masses comparable to that of the Sun, resulting from stellar evolution, it is now known that most galaxies harbor supermassive black

---

[10] Recall that the non-uniformity of a gravitational field is exactly what stops a free-falling frame from being globally equivalent to an inertial frame.

Fig. 6.8 Cygnus X-1 and Sagittarius A* (X-ray images by NASA's Chandra X-ray Observatory)

holes at their centers. The black hole at the center of our galaxy, *Sagittarius A**, weighs about 4.3 million solar masses and is about 26,000 light-years away.

Since black holes do not emit light, they cannot be directly observed. What is in fact observed are X-rays emitted by matter falling in (see Fig. 6.8).

6.8 Important Formulas

- Schwarzschild metric:

$$\Delta\tau^2 = \left(1 - \frac{2M}{r}\right)\Delta t^2 - \left(1 - \frac{2M}{r}\right)^{-1}\Delta r^2 - r^2\Delta\theta^2$$

- Gravitational acceleration measured by a stationary observer:

$$g = \frac{\frac{M}{r^2}}{\sqrt{1 - \frac{2M}{r}}}$$

- Proper time measured by a stationary observer:

$$\Delta\tau = \Delta t\sqrt{1 - \frac{2M}{r}}$$

- Redshift:

$$T' = T\sqrt{\frac{1 - \frac{2M}{r'}}{1 - \frac{2M}{r}}}$$

- Distance measured by a stationary observer:

$$\Delta s^2 = \left(1 - \frac{2M}{r}\right)^{-1} \Delta r^2 + r^2 \Delta\theta^2$$

- Precession of the perihelion:

$$\frac{6\pi M}{r} \text{ radians per orbit}$$

- Circular orbits:

$$\left(\frac{\Delta\theta}{\Delta t}\right)^2 = \frac{M}{r^3}$$

- Escape velocity for a stationary observer (see Exercise 5):

$$v = \sqrt{\frac{2M}{r}}$$

- Light deflection:

$$\frac{4M}{b} \text{ radians}$$

- Curvature of the Schwarzschild space-time (tidal force per unit mass per unit length):

$$\frac{M}{r^3}$$

6.9 Exercises

1. Show that if $\frac{M}{r}, \frac{M}{r'} \ll 1$ then the Schwarzschild gravitational redshift formula reduces to the approximate formula

$$T' = (1 + \Delta\phi)T.$$

2. Check that the precession of the perihelion of Mercury's orbit due to general relativistic effects is about 43 arcsec per century (distance from Mercury to the Sun: approximately 3.1 light-minutes). What is the precession of the perihelion of Earth's orbit due to these effects?

3. Compute the period of a circular orbit of radius $r > 3M$:

   (a) For an observer at infinity.
   (b) As seen by a stationary observer. What is the orbital velocity measured by these observers? What happens as r approaches $3M$?

Fig. 6.9 The sum of the
internal angles of an
Euclidean triangle is π

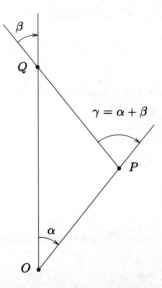

Fig. 6.10 Pulsar at the center
of the Crab nebula, about
6,500 light-years away
(X-ray image by NASA's
Chandra X-ray Observatory)

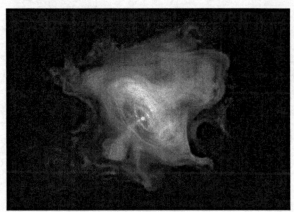

  (c) As measured by an orbiting observer. What happens as r approaches
      $3M$?
  (d) How is it possible that the free-falling observers in orbit measure a
      smaller period than the (accelerated) stationary observers?

4. (a) Show that the fact that the sum of the internal angles of an Euclidean
      triangle is π is equivalent to the statement that the angles α, β and γ in
      Fig. 6.9 satisfy $\alpha + \beta = \gamma$.
  (b) For speeds much smaller than the speed of light, the *angle* between two
      causal curves is just the relative velocity of the corresponding observers.
      Show that for these speeds the above relation is still valid in the Min-
      kowski geometry.

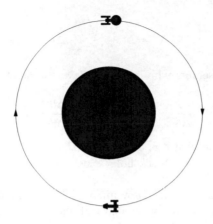

Fig. 6.11   Where Should Captain Kirk aim his Lasers?

      (c) Two circular orbits with the same radius r, traversed in opposite direc-
tions, form a two-sided polygon in the Schwarzschild geometry.
Approximating the polygon's area by half the area of the cylinder of
radius r and height equal to half the period of the orbit,[11] estimate the
curvature of the Schwarzschild solution for $\frac{M}{r} \ll 1$.

5. (a) What is the escape velocity for a stationary observer? What happens
when r approaches $2M$?

    (b) For which value of r is this velocity equal to the velocity of the circular
orbit? What is the corresponding velocity?

6. A typical *neutron star*[12] Fig. 6.10 has a mass of about 1.4 times the mass of
the Sun compressed into a sphere with a 10 km radius.

    (a) Compute the escape velocity for a stationary observer on the neutron
star's surface.

    (b) How many Earth gravities does the stationary observer measure?

7. Show that the deflection of a light ray just grazing the Sun's surface is about
1.75 arcsec (radius of the Sun: approximately 2.3 light-seconds).

8. Draw a space-time diagram describing the observation of the Einstein Cross
from Earth at a given time.

9. During an exciting space battle, the *Enterprise* and a Klingon warship fall into
the same circular orbit around a black hole, in diametrally opposite positions
(Fig. 6.11). Where should Captain Kirk aim his lasers?

10. Show that Painlevé's time coordinate t' satisfies

---

[11] Note that the map corresponding to the Schwarzschild coordinates preserves areas.

[12] Most neutron stars are rotating, so that their radio emissions are modulated into a periodic
signal with period equal to the period of rotation. Such neutron stars are called *pulsars*.

$$\Delta t' = \Delta t + \frac{\sqrt{\frac{2M}{r}}}{1 - \frac{2M}{r}} \Delta r.$$

11. Show that the observers who in the Painlevé coordinates satisfy

$$\frac{\Delta r}{\Delta t'} = -\sqrt{\frac{2M}{r}} \quad \text{and} \quad \frac{\Delta \theta}{\Delta t'} = 0$$

are free-falling, and that t' is their proper time.

12. A particle falling into a black hole is observed by a stationary observer at infinity. What does she see?

13. (a) What causes the tides? How many high tides are there per day?

   (b) Compute the approximate mass of the Moon from the fact that the Moon's tides are about twice as high as the Sun's tides.

14. Compute the radius of the event horizon of Sagittarius A^* in light-seconds and in solar radii.

15. Compute the tidal forces at the event horizon in Earth gravities per meter:

   (a) For Cygnus X-1.
   (b) For Sagittarius A^*.

6.10 Solutions

1. If $\frac{M}{r}, \frac{M}{r'} \ll 1$ then

$$\sqrt{\frac{1 - \frac{2M}{r'}}{1 - \frac{2M}{r}}} \simeq \sqrt{\left(1 - \frac{2M}{r'}\right)\left(1 + \frac{2M}{r}\right)} \simeq \sqrt{1 - \frac{2M}{r'} + \frac{2M}{r}} \simeq 1 - \frac{M}{r'} + \frac{M}{r}$$

$$= 1 + \phi' - \phi,$$

where ϕ and ϕ' are the gravitational potential at points with coordinates r and r'. Consequently,

$$T' = T\sqrt{\frac{1 - \frac{2M}{r'}}{1 - \frac{2M}{r}}} \simeq (1 + \Delta\phi)T.$$

2. The period T of a circular orbit of radius r satisfies

$$\frac{2\pi r}{T} = \sqrt{\frac{M}{r}} \Leftrightarrow T = 2\pi\sqrt{\frac{r^3}{M}}.$$

Therefore the period of Mercury's orbit is about

$$\sqrt{\left(\frac{3.1}{8.3}\right)^3} \simeq 0.23 \text{ years.}$$

The precession of Mercury's perihelion is about

$$\frac{6\pi \times 1.5}{3.1 \times 60 \times 300,000}$$

radians per orbit, that is, about

$$\frac{6\pi \times 1.5}{3.1 \times 60 \times 300,000} \times \frac{180}{\pi} \times 3,600 \times \frac{100}{0.23} \simeq 45$$

arcsec per century (the difference to the exact value of 43 arcsec per century is due to our approximations). The precession of Earth's perihelion is about

$$\frac{6\pi \times 1.5}{8.3 \times 60 \times 300,000}$$

radians per orbit, that is, about

$$\frac{6\pi \times 1.5}{8.3 \times 60 \times 300,000} \times \frac{180}{\pi} \times 3,600 \times 100 \simeq 4$$

arcseconds per century.

3. (a) Since circular orbits satisfy

$$\left(\frac{\Delta\theta}{\Delta t}\right)^2 = \frac{M}{r^3} \Leftrightarrow \Delta t = \pm\Delta\theta\sqrt{\frac{r^3}{M}},$$

we see that the period of the orbit for an observer at infinity (i.e. the value of Δt as $\Delta\theta = \pm 2\pi$) is

$$T_\infty = 2\pi\sqrt{\frac{r^3}{M}}.$$

(b) As seen by a stationary observer, the period is

$$T_S = T_\infty\sqrt{1 - \frac{2M}{r}} = 2\pi\sqrt{\frac{r^3}{M}}\sqrt{1 - \frac{2M}{r}}.$$

The orbital velocity measured by these observers is then

$$v = \frac{2\pi r}{T_S} = \frac{\sqrt{\frac{M}{r}}}{\sqrt{1 - \frac{2M}{r}}}.$$

As r approaches $3M$ this value tends to

$$\frac{\sqrt{\frac{1}{3}}}{\sqrt{1 - \frac{2}{3}}} = 1,$$

i.e. the speed of light. There are indeed null geodesics corresponding to light rays in circular orbits of radius $r = 3M$.

(c)   An orbiting observer satisfies $\Delta r = 0$ and

$$\Delta\theta^2 = \frac{M}{r^3}\Delta t^2,$$

and so she measures a proper time interval $\Delta\tau$ given by

$$\Delta\tau^2 = \left(1 - \frac{2M}{r}\right)\Delta t^2 - \left(1 - \frac{2M}{r}\right)^{-1}\Delta r^2 - r^2\Delta\theta^2$$

$$= \left(1 - \frac{2M}{r}\right)\Delta t^2 - \frac{M}{r}\Delta t^2$$

$$= \left(1 - \frac{3M}{r}\right)\Delta t^2,$$

that is,

$$\Delta\tau = \Delta t\sqrt{1 - \frac{3M}{r}}.$$

Consequently, the period of the orbit for an orbiting observer is

$$T_O = T_\infty\sqrt{1 - \frac{3M}{r}} = 2\pi\sqrt{\frac{r^3}{M}}\sqrt{1 - \frac{3M}{r}}.$$

Note that T_O tends to zero as r approaches $3M$. This was to be expected, as the velocity of the orbiting observer (with respect to the stationary observers) tends to the speed of light as r approaches $3M$.

(d)   Geometrically, the period measured by the orbiting observer is the length of the geodesic which represents a full orbit, whereas the period measured by the stationary observer is the length of a non-geodesic curve joining the same events (because the stationary observer is not free-falling). Therefore the geodesic corresponding to the circular orbit does not have maximum length. This phenomenon is due to the curvature of the Schwarzschild space-time. Analogously, there are always two geodesic segments connecting two non-antipodal points on the sphere, of which only one has minimum length (together they form the great circle defined by the two points). For the two events connected by the circular orbit, the maximizing geodesic is the one which describes a particle thrown upwards (i.e. in the radial direction) with the right speed so that it reaches its maximum altitude after half an orbit, returning to the initial point at the end of the orbit.

4. (a) The internal angles of the triangle are α, β and $\pi - \gamma$. Therefore we must have

$$\alpha + \beta + \pi - \gamma = \pi \Leftrightarrow \alpha + \beta = \gamma.$$

(b) With this interpretation, α is the velocity of OP with respect to OQ, β is the velocity of OQ with respect to PQ and γ is the velocity of OP with respect to PQ. Since the velocities are much smaller than the speed of light we have $\gamma = \alpha + \beta$.

(c) For $\frac{M}{r} \ll 1$, the speed of each of the circular orbits is approximately

$$v = \sqrt{\frac{M}{r}}.$$

Each of the polygon's angles is approximately $2v$, and the sum of the polygon's internal angles is then approximately $4v$. The area of the cylinder of radius r and height equal to half the orbit's period is approximately

$$A = 2\pi r \times \frac{\pi r}{v} = \frac{2\pi^2 r^2}{v}.$$

Therefore the curvature of the Schwarzschild solution for $\frac{M}{r} \ll 1$ is of the order

$$\frac{8v}{A} = \frac{8v^2}{2\pi^2 r^2} = \frac{4}{\pi^2}\frac{M}{r^3}.$$

5. (a) A particle thrown in the radial direction has angular momentum

$$L = r^2 \frac{\Delta\theta}{\Delta\tau} = 0,$$

and so the differential equations describing its motion are

$$\frac{\Delta r}{\Delta\tau} = \pm\sqrt{2E + \frac{2M}{r}};$$

$$\frac{\Delta t}{\Delta\tau} = \left(1 - \frac{2M}{r}\right)^{-1}\sqrt{1 + 2E}.$$

The first equation implies

$$2E + \frac{2M}{r} \geq 0 \Leftrightarrow -Er \leq M.$$

If $E < 0$ then the range of the r coordinate is limited. Consequently the escape velocity corresponds to $E = 0$. The above equations then imply

$$\frac{\Delta r}{\Delta t} = \sqrt{\frac{2M}{r}}\left(1 - \frac{2M}{r}\right).$$

The distance measured by a stationary observer is

$$\Delta s = \left(1 - \frac{2M}{r}\right)^{-\frac{1}{2}} \Delta r,$$

and her proper time is

$$\Delta \tau = \Delta t \sqrt{1 - \frac{2M}{r}}.$$

Consequently the escape velocity measured by a stationary observer is

$$v = \frac{\Delta s}{\Delta \tau} = \left(1 - \frac{2M}{r}\right)^{-1} \frac{\Delta r}{\Delta t} = \sqrt{\frac{2M}{r}}$$

(which coincidentally is the Newtonian result). Note that when r approaches $2M$ the escape velocity approaches 1 (i.e. the speed of light).

(b) Recall from Exercise 3 that the orbital speed measured by a stationary observer is

$$v = \frac{\sqrt{\dfrac{M}{r}}}{\sqrt{1 - \dfrac{2M}{r}}}.$$

This speed is equal to the escape velocity for

$$\frac{\sqrt{\dfrac{M}{r}}}{\sqrt{1 - \dfrac{2M}{r}}} = \sqrt{\frac{2M}{r}} \Leftrightarrow 1 = 2\left(1 - \frac{2M}{r}\right) \Leftrightarrow r = 4M,$$

corresponding to a velocity $\frac{\sqrt{2}}{2} \simeq 71\%$ of the speed of light. The circular orbit of radius $r = 4M$, like all orbits with radius smaller than $6M$, is unstable, and can be reached by a particle dropped from infinity with the right angular momentum.

6. (a) The escape velocity for a stationary observer on the neutron star's surface is

$$v \simeq \sqrt{\frac{2 \times 1.4 \times 1.5}{10}} \simeq 0.65$$

Fig. 6.12 Space-time
diagram describing the
observation of the Einstein
Cross from Earth

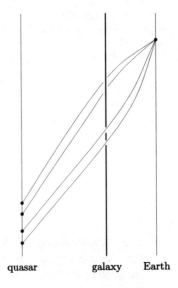

quasar   galaxy   Earth

(i.e. about 65% of the speed of light).

(b) Using km as units, the gravitational field measured by the stationary
observer on the neutron star's surface is

$$g = \frac{\frac{M}{r^2}}{\sqrt{1 - \frac{2M}{r}}} \simeq \frac{\frac{1.4 \times 1.5}{10^2}}{\sqrt{1 - \frac{2 \times 1.4 \times 1.5}{10}}} \simeq 2.8 \times 10^{-2}.$$

In the same units, the Earth's gravitational acceleration is

$$g = \frac{M}{r^2} \simeq \frac{4.5 \times 10^{-6}}{6,400^2} \simeq 1.1 \times 10^{-13}.$$

Therefore the gravitational field measured by the stationary observer on the
neutron star's surface is about

$$\frac{2.8 \times 10^{-2}}{1.1 \times 10^{-13}} \simeq 2.5 \times 10^{11}$$

Earth gravities.

7. Using km as units, we see that the deflection of a light ray just grazing the
Sun's surface is about

$$\frac{4 \times 1.5}{2.3 \times 300,000} \times \frac{180}{\pi} \times 3,600 \simeq 1.8 \, \text{arcsec}$$

(the difference to the exact value of 1.75 arcsec is due to our approximations).

Fig. 6.13 Captain Kirk
should aim his lasers in the
same angle as he sees the
Klingon warship

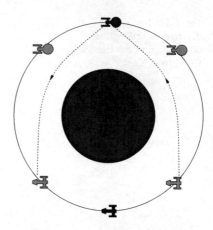

8. The space-time diagram describing the observation of the Einstein Cross from
   Earth at a given time is depicted in Fig. 6.12. There are four null geodesics
   connecting the history of the quasar to the event in which it is observed on
   Earth at the given time. Generically, the light rays corresponding to these
   geodesics were emitted at different times.
9. There are two factors complicating Captain Kirk's aim: the light deflection by
   the black hole, which makes the laser beam follow a curved trajectory, and the
   fact that light propagates with finite speed, which forces Captain Kirk to aim
   not at the current position of the Klingon warship but to where it will be when
   the laser beam reaches it. However, the same factors affect the light that the
   Enterprise is receiving from the Klingon warship. By symmetry, it is then
   clear that Captain Kirk should aim his lasers in the same angle as he sees the
   Klingon warship, but in the opposite direction, that is, in the direction that the
   light from the Klingon warship would go if it were reflected by the *Enterprise*
   (Fig. 6.13).
10. We just have to note that with this definition the Schwarzschild metric
    becomes

$$\Delta\tau^2 = \left(1 - \frac{2M}{r}\right)\left(\Delta t' - \frac{\sqrt{\frac{2M}{r}}}{1 - \frac{2M}{r}}\Delta r\right)^2 - \left(1 - \frac{2M}{r}\right)^{-1}\Delta r^2 - r^2\Delta\theta^2$$

$$= \left(1 - \frac{2M}{r}\right)\Delta t'^2 - 2\sqrt{\frac{2M}{r}}\Delta t'\Delta r - \Delta r^2 - r^2\Delta\theta^2.$$

11. In Painlevé coordinates we have

$$\Delta\tau = \Delta t'\sqrt{1 - \left(\frac{\Delta r}{\Delta t'} + \sqrt{\frac{2M}{r}}\right)^2 - r^2\left(\frac{\Delta\theta}{\Delta t'}\right)^2},$$

and so for observers satisfying

Fig. 6.14 A particle falling into a *black hole* is observed by a stationary observer

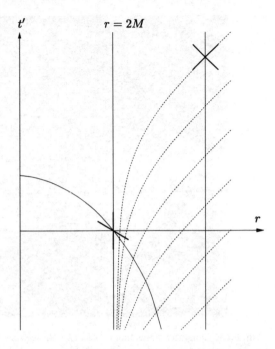

$$\frac{\Delta r}{\Delta t'} = -\sqrt{\frac{2M}{r}} \quad \text{and} \quad \frac{\Delta \theta}{\Delta t'} = 0$$

we have $\Delta \tau = \Delta t'$. Let P and Q be two events in the history γ of one of these observers, with Painlevé time coordinates t'_P and t'_Q. Then the length of γ between P and Q is $t'_Q - t'_P$. For any other causal curve γ' connecting the same events we clearly have $\Delta \tau \leq \Delta t'$, and so the length of γ' is less than or equal to $t'_Q - t'_P$. We conclude that γ is a causal curve of maximum length, that is, a geodesic. Therefore γ is the history of a free-falling observer.

12. The light received by the stationary observer at infinity corresponds to the null geodesics moving away from the event horizon. These geodesics cannot cross the horizon, and so they accumulate along it, as shown in Fig. 6.14. Consequently all of them intersect the history of the falling particle. This means that the observer at infinity never stops seeing the particle, although she sees it moving slower and slower, as if suspended just outside the horizon, more and more redshifted. (This is true even for the particles which collapsed to form the black hole in the first place: as seen by an observer at infinity, they never actually cross the horizon! In practice, however, the particles quickly become too redshifted to be detectable, and so the black hole is seen as a dark sphere against a background distorted by strong gravitational lensing, as shown in Fig. 6.15).

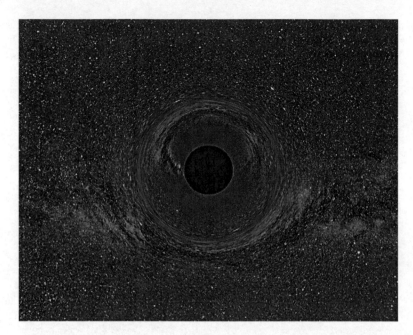

Fig. 6.15 Computer simulation of a *black hole* seen against the Milky Way. (Image credit: Ute Kraus, Institut für Physik und Technik, Universität Hildesheim, Germany)

13. (a) The Earth is free-falling in the Moon's gravitational field. But the Earth has non-negligible dimensions. So it is actually the center of the Earth which is free-falling. The regions of the Earth closer to the Moon are being more strongly attracted than the center of the Earth, and the regions further from the Moon are being less strongly attracted. This creates a residual stretching force along the Earth-Moon axis. This effect is more noticeable on the oceans, because they are more deformable than the crust, and so their level is higher along the Earth-Moon axis. As the Earth rotates, these two higher level regions move across the Earth's surface. Therefore there are *two* high tides per day. The Sun also gives rise to (lower) tides. The relative position of the Sun and the Moon determines the intensity of the tides: they are higher when the Sun and the Moon are aligned (i.e. during a new Moon or a full Moon).

(b) Let m be the mass of the Moon and M the mass of the Sun. We have

$$\frac{m}{1.3^3} \simeq 2 \times \frac{M}{(8.3 \times 60)^3} \Leftrightarrow m \simeq 3.6 \times 10^{-8} \times M \simeq 5.4 \times 10^{-5}\,\text{m}.$$

So the Earth is about $\dfrac{4.5 \times 10^{-3}}{5.4 \times 10^{-5}} \simeq 83$ times more massive than the Moon.

14. Since the mass of Sagittarius A* is about 4.3 million solar masses, the radius of its event horizon is about

$$2 \times 4.3 \times 10^6 \times 1.5 = 4.3 \times 3 \times 10^6 \, \text{km},$$

that is, about 43 light-seconds, corresponding to approximately $\dfrac{43}{2.3} \simeq 19$ solar radii.

15. (a) Using meters as units, the Earth's gravitational acceleration is

$$g = \frac{4.5 \times 10^{-3}}{(6.4 \times 10^6)^2}.$$

So the tidal forces at the Cygnus X-1 event horizon are approximately

$$\frac{9 \times 1.5 \times 10^3}{(2 \times 9 \times 1.5 \times 10^3)^3} \times \frac{(6.4 \times 10^6)^2}{4.5 \times 10^{-3}} \simeq 6.2 \times 10^6 \, \text{gravities per meter}$$

(that is, two points 1 meter apart experience a difference of gravitational acceleration of 6.2 million Earth gravities). So an astronaut falling into this black hole would be crushed much before reaching the horizon.

(b) Similarly, the tidal forces at the Sagittarius A* event horizon are approximately

$$\frac{4.3 \times 10^6 \times 1.5 \times 10^3}{(2 \times 4.3 \times 10^6 \times 1.5 \times 10^3)^3} \times \frac{(6.4 \times 10^6)^2}{4.5 \times 10^{-3}} \simeq 2.7 \times 10^{-5} \, \text{gravities per meter}.$$

So an astronaut falling into this black hole would hardly notice the tidal forces as she crossed the horizon.

Chapter 7
Cosmology

In this chapter we discuss the grandest subject of all—cosmology, the study of the Universe as a whole. After defining what is meant by the redshift of a light signal, we learn how Hubble, having realized that the Milky Way is just one of an immense number of galaxies, went on to discover that most of them are moving away from us, with velocities increasing linearly with distance (Hubble law). We introduce the Friedmann-Lemaître-Robertson-Walker (FLRW) models for the Universe, which are the only space–time metrics for which space is homogeneous (the same everywhere) and isotropic (the same in every direction). We see how these models explain the Hubble law, and exactly what the redshift of a distant galaxy means. Finally, we analyze the consequences of the Einstein equation, which in the FLRW models reduces to the Friedmann equations for the density and radius of the Universe. We see how these equations imply that the Universe originated in a Big Bang, and will, according to the currently accepted cosmological parameters, expand forever.

7.1 Redshift

Consider a light signal emitted with period T. If a given observer measures a period T' then the signal is said to have suffered a *redshift*

$$z = \frac{T' - T}{T} = \frac{T'}{T} - 1.$$

Equivalently,

$$T' = T(1 + z).$$

From Exercise 5 in Chap. 2 we know that, in the absence of gravity and for low velocities (as compared to the speed of light), the redshift z coincides with the

J. Natário, *General Relativity Without Calculus*, Undergraduate Lecture Notes in Physics, DOI: 10.1007/978-3-642-21452-3_7,
© Springer-Verlag Berlin Heidelberg 2011

velocity v at which the signal's source is receding. In Chap. 5 we also saw that stationary observers in weak gravitational fields measure a redshift $z = \Delta\phi$, where $\Delta\phi$ is the gravitational potential difference between the observer and the source.

The atoms of a given element (say hydrogen) can only absorb radiation with certain well defined periods. After crossing a region containing an appreciable amount of atoms of such an element, the spectrum of white light displays a series of dark absorption lines, which constitute the signature of that element. When the light is received on Earth these lines are in general shifted with respect to their usual positions; measuring the displacement one obtains the redshift and, consequently, the velocity of the absorbing atoms. That is how astronomers measure the velocity at which a given star recedes from Earth. (There is also a small shift due to the star's gravity, but it is much smaller than the velocity effect—see Exercise 1).

7.2 Hubble Law

The existence of other galaxies besides ours (the Milky Way) was only established in 1924 by Hubble[1] (Fig. 7.1). As a follow-up to his work, Hubble carefully measured the redshift of about 50 galaxies, and discovered that not only were most of them moving away from us but also that the greater the distance the greater the recession speed. In 1929 Hubble formulated what is nowadays called the *Hubble law*: $v = Hd$, where v is the recession speed of a given galaxy, d is its distance from Earth and H is a constant, called the *Hubble constant*, whose modern value is about 71 km/s per megaparsec (a *parsec* is a length unit approximately equal to 3.26 light-years—see Exercise 2; a *megaparsec* is a million parsecs). In other words, a galaxy 1 megaparsec away is in average receding at 71 km/s. For comparison's sake, the Milky Way's diameter is about 0.03 megaparsecs, the Andromeda and Triangulum galaxies (which, together with the Milky Way, are the most important galaxies of the so-called Local Group) are about 0.8 megaparsecs away, and the radius of the observable Universe is about 14,000 megaparsecs.

7.3 FLRW Models

The first space–time describing the Universe on large scales (usually called a *cosmological model*) was proposed by Einstein in 1917. Einstein started from the hypothesis that the Universe is a *hypersphere*, that is, the three-dimensional analogue of the (two-dimensional) surface of a sphere (the thought that space was finite in extent but without boundary was philosophically appealing to him).

---

[1] Edwin Hubble (1889–1953), American astronomer.

Fig. 7.1 Edwin Hubble
(reproduced by permission of
the Huntington Library, San
Marino, California)

Introducing this hypothesis in his equation, Einstein discovered that, as he expected, the Universe's density would have to be constant. This is not unreasonable: although on our scale the Universe is not homogeneous, since matter is clumped into stars and stars into galaxies, at larger scales these fine details can be ignored. To his surprise, however, Einstein discovered that such an Universe would not remain static: its radius would decrease until the hypersphere collapsed to a point. To keep his Universe static, Einstein was forced to introduce a *cosmological constant* (sometimes also called *dark energy*), representing the energy of the vacuum, whose gravitational effect would be repulsive. Later, when the expansion of the Universe was discovered (thus showing that the Universe is not static), Einstein considered the introduction of this constant "his greatest blunder". In 1998, however, astronomers found irrefutable evidence that the cosmological constant does in fact exist.

Still in 1917, de Sitter[2] (Fig. 7.2) presented another cosmological model, representing a hyperspherical Universe without matter but with a positive cosmological constant. Actually, both the Einstein and the de Sitter Universes are particular cases of the general models discovered by Friedmann[3] in 1922, and rediscovered by Lemaître[4] in 1927, and by Robertson[5] and Walker[6] in 1929. The Friedmann-Lemaître-Robertson-Walker (FLRW) models assume only that the Universe on large scales is *homogeneous* (i.e. the same at all points) and *isotropic* (i.e. the same in all directions); such hypotheses were spectacularly confirmed in

---

[2] Willem de Sitter (1872–1934), Dutch mathematician, physicist and astronomer.

[3] Alexander Friedmann (1888–1925), Russian mathematician and meteorologist.

[4] Georges-Henri Lemaître (1894–1966), Belgian Catholic priest and astronomer.

[5] Howard Robertson (1903–1961), American mathematician and physicist.

[6] Arthur Walker (1909–2001), English mathematician.

Fig. 7.2 de Sitter, Friedmann, Lemaître, Robertson and Walker

Fig. 7.3 Alpher, Gamow, Herman, Penzias and Wilson

1964, with the discovery of the *cosmic background radiation*, by Penzias[7] and Wilson[8] (Fig. 7.3). This radiation, predicted in 1948 by Alpher,[9] Gamow[10] and Herman,[11] consists of microwave photons (about 400 per cubic cm) which are thought to have been produced in the primitive Universe; they come from all directions in the sky and have exactly the same spectrum, independently of the direction which they are coming from.

There are only three kinds of spaces which are homogeneous and isotropic: the hypersphere, the usual Euclidean space (on which the Pythagorean theorem holds) and the *hyperbolic space*. The hyperbolic space, which, like the Euclidean space, is infinite in extent, is a kind of opposite of the hypersphere, with constant negative curvature: the sum of the internal angles of a triangle in this space is always *less* than π. By analogy with the hypersphere, one defines the radius of the hyperbolic space as the number R such that its curvature is $-\frac{1}{R^2}$.

Since the FLRW models require that the space is either the hypersphere, the Euclidean space or the hyperbolic space, the only thing that can change in time is the Universe's radius R. In this way, and restricting ourselves as usual to two spatial dimensions, the metric of the FLRW models is given by

[7] Arno Penzias (1933–), American physicist, winner of the Nobel prize in physics (1978).

[8] Robert Wilson (1936–), American physicist, winner of the Nobel prize in physics (1978).

[9] Ralph Alpher (1921–2007), American physicist.

[10] George Gamow (1904–1968), Ukrainian physicist.

[11] Robert Herman (1914–1997), American physicist.

$$\Delta\tau^2 = \Delta t^2 - R(t)^2 \left(\Delta\theta^2 + \cos^2\theta\Delta\varphi^2\right) \qquad \text{or}$$
$$\Delta\tau^2 = \Delta t^2 - R(t)^2 \left(\Delta x^2 + \Delta y^2\right) \qquad \text{or}$$
$$\Delta\tau^2 = \Delta t^2 - R(t)^2 \left(\Delta\theta^2 + \cosh^2\theta\Delta\varphi^2\right)$$

according to whether space is hyperspherical, Euclidean or hyperbolic (the exact form of the function cosh, called the *hyperbolic cosine*, will not be needed).

7.4 Hubble Law in the FLRW Models

In problems for which only one spatial dimension is relevant we may restrict the FLRW metric to a line where φ (or y in the Euclidean case) is constant. In this way, the metric becomes the same for the three models:

$$\Delta\tau^2 = \Delta t^2 - R(t)^2 \Delta\theta^2$$

(where we have set $x = \theta$ in the Euclidean case).

The lines where the coordinate θ is kept constant are geodesics: their length between two events P and Q with time coordinates t_P and t_Q is $t_Q - t_P$, whereas any other causal curve connecting the same events satisfies

$$\Delta\tau = \sqrt{\Delta t^2 - R(t)^2 \Delta\theta^2} \leq \Delta t,$$

and so its length is less than or equal to $t_Q - t_P$. The Einstein equation implies that these geodesics are the histories of the matter particles (galaxies), which are then free-falling. Moreover, the time coordinate t is simply the proper time measured by the galaxies.

If two galaxies are located at positions $\theta = \theta_1$ and $\theta = \theta_2$ then the distance between them at time t is

$$d(t) = R(t)(\theta_2 - \theta_1).$$

Consequently, in a time interval Δt the distance varies by

$$\Delta d = \Delta R(\theta_2 - \theta_1),$$

and so the galaxies move away at a velocity

$$v = \frac{\Delta d}{\Delta t} = \frac{\Delta R}{\Delta t}(\theta_2 - \theta_1) = \frac{\Delta R}{\Delta t}\frac{d}{R} = Hd,$$

where we have defined

$$H(t) = \frac{1}{R}\frac{\Delta R}{\Delta t}.$$

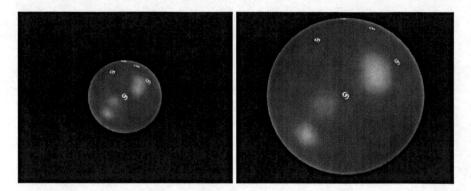

Fig. 7.4 The points on the surface of a balloon which is being inflated behave like the galaxies in the expanding Universe. (Image credit: Rob Knop, Quest University, Canada)

Thus we see that the the FLRW models explain the Hubble law naturally as a consequence of the variation of the radius of the Universe (note however that in general the Hubble constant varies in time). The situation is analogue to what happens to the points on the surface of a balloon which is being inflated: each point sees the other points moving away with a velocity proportional to their distance. One should bear in mind, however, that this is just an analogy. In particular, in the case of the Universe there is nothing that corresponds to the interior (or the exterior) of the balloon (Fig. 7.4).

7.5 Redshift in the FLRW Models

The FLRW metric is invariant under translations in θ. This means that the interval between nearby events P_1 e P_2, with coordinates (t_1, θ_1) and (t_2, θ_2), is equal to the interval between the events Q_1 and Q_2 with coordinates $(t_1, \theta_1 + \Omega)$ and $(t_2, \theta_2 + \Omega)$, for any $\Omega \in \mathbb{R}$. So if γ is a geodesic, that is, a curve of maximum length, the curve γ' obtained from γ by moving all of its points to points with the coordinate θ increased by Ω is also a geodesic. The same is true if γ is a null geodesic.

Consider two galaxies O and O', with coordinates θ and θ'. Suppose that O sends a light signal with period T towards O' at time t. The histories of the light rays corresponding to the beginning and to the end of the period are null geodesics γ and γ', and so they satisfy

$$\Delta t^2 - R(t)^2 \Delta \theta^2 = 0 \Leftrightarrow \frac{\Delta t}{\Delta \theta} = R(t)$$

(assuming that the signal is propagating in the direction of increasing θ). From Fig. 7.5 it is easy to see that γ can be obtained from γ' by a translation of Ω in the coordinate θ, and also that the period T measured by O is related to Ω by

Fig. 7.5 Redshift in the
FLRW geometry

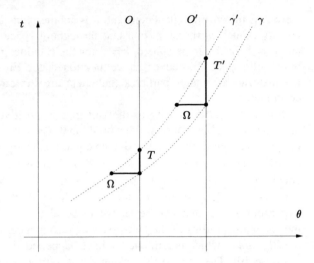

$$\frac{T}{\Omega} = \frac{\Delta t}{\Delta \theta} = R(t).$$

Similarly, the period T' measured by O' satisfies

$$\frac{T'}{\Omega} = \frac{\Delta t}{\Delta \theta} = R(t'),$$

where t' is the time at which the signal is received. We conclude that

$$1 + z = \frac{T'}{T} = \frac{R'}{R},$$

where R is the radius of the Universe at the time of emission and R' is the radius of
the Universe at the time of reception. Thus the redshift z does not yield directly the
velocity v at which the source is receding; instead, it measures how much the
Universe has expanded since the light signal was emitted. For small redshifts,
however, one has $z \simeq v$ (see Exercise 6).

7.6 Friedmann Equations

The variation of $R(t)$ is determined by the Einstein equation, which in the FLRW
models reduces to the *Friedmann equations*:

$$\frac{\Delta R}{\Delta t} = \pm \sqrt{\frac{2E}{R} + \frac{\Lambda R^2}{3} - k};$$

$$E = \frac{4\pi R^3}{3} \rho.$$

Here Λ is the cosmological constant, E is another constant, ρ is the average matter density (which is therefore constant throughout space, depending only on time) and $k = 1$ for the hypersphere, $k = 0$ for the Euclidean space and $k = -1$ for the hyperbolic space. Moreover, as we mentioned, the Einstein equation implies that the histories of matter particles (galaxies) are the curves with constant spatial coordinates.

Since the matter density of the Universe is positive, we have $E > 0$. On the other hand, Hubble's observations imply that R is increasing, and so we must take the positive sign in the first Friedmann equation. Moreover, it is known since 1998 (from supernovae observations) that $\frac{\Delta R}{\Delta t}$ is increasing; from the first Friedmann equation it is then clear that we must have $\Lambda > 0$. Consequently, the radius of the Universe is strictly increasing for all time (this is obvious in the Euclidean and hyperbolic cases, and can be shown to be also true for the hypersphere). This means that the radius of the Universe was zero at a certain point in its past (called the *Big Bang*), which is estimated to have happened about 14 billion years ago (see Exercise 10). The second Friedmann equation implies that the Universe's matter density was infinite then (as well as its temperature—see Exercise 16). Moreover, we see that the fate of the Universe is to expand forever (at an increasing rate), becoming ever colder and sparser.

The first Friedmann equation can be rewritten as

$$H^2 = \frac{8\pi}{3}(\rho + \rho_\Lambda) - \frac{k}{R^2},$$

where

$$\rho_\Lambda = \frac{\Lambda}{8\pi}$$

represents the vacuum energy density. We then see that the Universe is hyperspherical ($k = 1$), Euclidean ($k = 0$) or hyperbolic ($k = -1$) according to whether the Universe's total density $\rho + \rho_\Lambda$ is greater than, equal to or less than the *critical density*[12]

$$\rho_c = \frac{3H^2}{8\pi}.$$

Unfortunately, current observations are not accurate enough to decide which of these cases actually occurs. It is known however that the Universe is very close to being Euclidean, with

$$\frac{\rho}{\rho_c} \simeq 27\% \quad \text{and} \quad \frac{\rho_\Lambda}{\rho_c} \simeq 73\%.$$

---

[12] Because of Einstein's mass-energy equivalence relation $E = mc^2$, which in our units is written $E = m$, we do not distinguish between mass density and energy density.

The matter which is directly observable—stars, galaxies—is only 20% of the total amount of matter which must exist to explain the motions of stars and galaxies. The remaining 80% correspond to the so-called *dark matter* (not to be confused with dark energy), whose exact nature is not yet known.

7.7 Important Formulas

- Redshift:

$$\boxed{T' = T(1+z)}$$

- Hubble law:

$$\boxed{v = Hd} \quad \text{with} \quad \boxed{H = \frac{1}{R}\frac{\Delta R}{\Delta t}}$$

- Redshift in the FLRW models:

$$\boxed{1 + z = \frac{R'}{R}}$$

- First Friedmann equation:

$$\boxed{\frac{\Delta R}{\Delta t} = \pm\sqrt{\frac{2E}{R} + \frac{\Lambda R^2}{3} - k}} \quad \text{or} \quad \boxed{H^2 = \frac{8\pi}{3}(\rho + \rho_\Lambda) - \frac{k}{R^2}}$$

- Second Friedmann equation:

$$\boxed{E = \frac{4\pi R^3}{3}\rho}$$

- Vacuum energy density:

$$\boxed{\rho_\Lambda = \frac{\Lambda}{8\pi}}$$

- Critical density:

$$\boxed{\rho_c = \frac{3H^2}{8\pi}}$$

- Observed densities:

$$\boxed{\frac{\rho}{\rho_c} \simeq 27\%} \quad \text{and} \quad \boxed{\frac{\rho_\Lambda}{\rho_c} \simeq 73\%}$$

7.8 Exercises

1. What is the speed corresponding to the gravitational redshift of light emitted from the Sun's surface?
2. The *parallax* of a star is *half* the maximum shift of its apparent position in the sky due to the Earth's yearly motion. A *parsec* is the distance to a star whose parallax is 1 arcsec. Show that 1 parsec is approximately 3.26 light-years.
3. The Sun orbits the center of the Milky Way at a distance of about 8 kpc, with a speed of about 220 km/s. What is the period of its orbit? How many orbits has the Sun completed since its formation, about 4.5 billion years ago?
4. What should the velocity of the Andromeda galaxy with respect to the Earth be according to the Hubble law?
5. What is the recession speed of:

    (a) The remains of the most distant supernova ever detected (SN 19941), about 5,500 megaparsecs from Earth?
    (b) The most distant known quasar (SDSS J1030 + 0524), about 8,500 megaparsecs from Earth?
    (c) A galaxy at the boundary of the observable Universe, about 14,000 megaparsecs from Earth? Does this contradict the fact that nothing can move faster than the speed of light?

6. Show that for small redshifts one has $z \simeq v$.
7. For how long can be seen on Earth a phenomenon which lasts 10 days:

    (a) On the most distant supernova ever detected (SN 19941), whose redshift is $z \simeq 2.4$?
    (b) On the most distant known quasar (SDSS J1030 + 0524), whose redshift is $z \simeq 6.3$?

8. The observed brightness of a galaxy is determined by its light flux, that is, the amount of energy emitted by the galaxy which crosses a unit area per unit time on Earth. If the galaxy is at a distance d from Earth then its light will be spread over the surface of a sphere of radius d. So the light flux in an Euclidean Universe is

$$F = \frac{L}{4\pi d^2},$$

where L is the galaxy's luminosity, that is, the total amount of energy that it emits per unit time.

    (a) If the redshift is not negligible then this formula must be corrected to

$$F = \frac{L}{4\pi d^2 (1 + z)^2}.$$

Why?

(b) Is the flux at the same distance d larger or smaller in an hyperspherical Universe?

9. The current cosmological observations imply

$$\frac{\rho + \rho_\Lambda}{\rho_c} < 1.01.$$

Show that if the Universe is hyperspherical then its radius is larger than 42,000 megaparsecs. If this were indeed the radius of the Universe what would be the distance to the antipodal point of Earth? How fast would it be receding?

10. Estimate the age of the Universe assuming $\frac{\Delta R}{\Delta t}$ constant.

11. For a long time it was thought that the cosmological constant was exactly zero. How would the Universe evolve in this case?

12. Show that a free-falling particle in the Einstein Universe moves along a geodesic of the hypersphere. What would an observer see in the Einstein Universe?

13. All objects visible from Earth are at a distance smaller than about 14,000 megaparsecs (radius of the *observable* Universe). It is said that the points which are at this distance from Earth form its *cosmological horizon*.

(a) Why is there a cosmological horizon?

(b) What is the redshift of an object at the horizon?

(c) How long would light take to travel the distance to the horizon?

(d) Compare this time interval with the age of the Universe. Is there a contradiction here?

14. Show that:

(a) The Hubble constant will always be larger than 61 km/s/Mpc.

(b) The radius of the Universe will more than double every 16 billion years.

(c) Light being emitted now by a galaxy more than 32 billion light-years away will never reach the Earth.

(d) Objects more than 24,000 megaparsecs away will never be seen from Earth.

15. Why is the sky dark at night? Is the sky dark at night in the Einstein Universe?

16. The temperature of a system is a measure of the average energy per particle.

(a) Show that the temperature of the cosmic background radiation is inversely proportional to the radius of the Universe. What was its temperature at the Big Bang?

Fig. 7.6 The *Enterprise* crosses a wormhole into an unknown region. (STAR TREK and related marks are trademarks of CBS Studios Inc.)

(b) It is known that the cosmic background radiation ceased to interact with matter when its temperature was about 2,940 K, and that its current temperature is about 2.7 K. How much redshift has the cosmic background radiation suffered?

(By definition water freezes at 273 K and boils at 373 K; the zero of this scale is called the *absolute zero*, because it is the lowest attainable temperature).

17. The *Enterprise* crosses a wormhole into an unknown region of space–time (Fig. 7.6). The astrophysical section reports measuring a cosmic background radiation temperature of 5.4 K and a Hubble constant of 90 km/s/Mpc. Is the *Enterprise* in our Universe?

7.9 Solutions

1. The gravitational redshift of light emitted from the Sun's surface is equal to the gravitational potential difference between infinity and the Sun's surface, that is, about $\frac{1.5}{2.3 \times 300,000}$. This is the redshift due to a speed of about $300,000 \times \frac{1.5}{2.3 \times 300,000} \simeq 0.7$ km/s. For comparison's sake, the speeds of nearby stars with respect to the Sun are of the order of tens (or even hundreds) of km per second.

2. Figure 7.7 depicts the fact that a star's parallax is the angle α between the line connecting the Sun to the star and the line connecting the Earth to the star. This angle is usually very small, and so approximately equal to its tangent

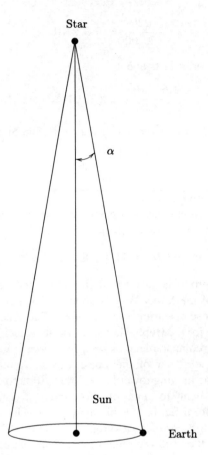

Fig. 7.7 Parallax of a star

(when expressed in radians). So the distance from the Sun to the star when the parallax is 1 arcsec is approximately

$$\frac{8.3}{\pi/(180 \times 3,600)} \text{ light-minutes,}$$

that is, about

$$\frac{8.3}{\pi/(180 \times 3,600) \times 60 \times 24 \times 365} \simeq 3.26 \text{ light-years.}$$

3. To complete an orbit the Sun must travel about

$$2\pi \times 8 \times 10^3 \times 3.26 \simeq 1.6 \times 10^5 \text{ light-years.}$$

Since its speed is about

$$\frac{220}{300,000} \simeq 7.3 \times 10^{-4},$$

we see that each orbit lasts about

$$\frac{1.6 \times 10^5}{7.3 \times 10^{-4}} \simeq 2.2 \times 10^8 \text{ years,}$$

that is, about 220 million years. Consequently, the Sun has completed

$$\frac{4.5}{0.22} \simeq 20$$

orbits since its formation.

4. According to the Hubble law the Andromeda galaxy should be moving away from the Earth at a velocity of about

$$v = Hd \simeq 71 \times 0.8 = 57 \text{ km/s.}$$

Actually, the Andromeda galaxy is *approaching* the Earth at about 300 km/s (and the center of the Milky Way at about 130 km/s), thus being one of the few galaxies whose spectrum is *blueshifted*. The Andromeda galaxy is too close to the Earth for the Hubble law to work: its velocity due to the expansion of the Universe is comparable to its *peculiar velocity*, resulting from its orbital motion around the center of the Local Group. Actually, the whole Local Group seems to be moving towards the Great Attractor, a large concentration of matter equivalent to tens of thousands of galaxies, located about 75 megaparsecs form Earth. For this reason, the velocity of the Local Group with respect to the cosmic background radiation (which defines the best approximation to the idealized reference frame of the FLRW models) is about 600 km/s (the Sun moves at about 380 km/s in this frame).

5. (a) According to the Hubble law, the remains of the most distant supernova ever detected are moving away from the Earth at a speed of about

$$v = Hd \simeq 71 \times 5,500 \simeq 390,000 \text{ km/s,}$$

   that is, about $\frac{390,000}{300,000} = 1.3$ times the speed of light.

   (b) According to the Hubble law, the most distant known quasar is moving away from Earth at a speed of about

$$v = Hd \simeq 71 \times 8,500 \simeq 600,000 \text{ km/s,}$$

   that is, about $\frac{600,000}{300,000} = 2$ times the speed of light.

   (c) According to the Hubble law, a galaxy at the boundary of the observable Universe is moving away from Earth at a speed of about

$$v = Hd \simeq 71 \times 14{,}000 \simeq 990{,}000 \text{ km/s,}$$

that is, about $\dfrac{990{,}000}{300{,}000} = 3.3$ times the speed of light.

There is no contradiction here, since the histories of these objects are causal curves: they do not move faster than any light signal in their neighborhood. What happens is that in a given time interval the distance between them and the Earth increases more than the distance that light would travel in the same time interval. In the balloon analogy, what is forbidden is for objects to move faster than the speed of light with respect to the balloon's surface; if the balloon is inflated fast enough, however, the distance between two points on its surface can increase arbitrarily fast.

6. The Hubble law is

$$v = Hd = \frac{1}{R}\frac{\Delta R}{\Delta t}d.$$

If $z \ll 1$ then the radius of the Universe does not increase much in the time interval Δt that light takes to travel the distance d, and so $d \simeq \Delta t$. Consequently we obtain the approximate formula

$$v \simeq \frac{\Delta R}{R}.$$

Since

$$z = \frac{R'}{R} - 1 = \frac{R' - R}{R} = \frac{\Delta R}{R},$$

we see that for $z \ll 1$ one has

$$z \simeq v.$$

7. (a) A phenomenon which lasts 10 days on the supernova can be seen on Earth during

$$T' = T(1 + z) \simeq 10 \times (1 + 2.4) = 34 \text{ days.}$$

This effect has been observed for several supernovae (whose average duration is known).

   (b) A phenomenon which lasts 10 days on the quasar can be seen on Earth during

$$T' = T(1 + z) \simeq 10 \times (1 + 6.3) = 73 \text{ days.}$$

8. (a) Due to the redshift, photons emitted by the galaxy during a time interval
        T are observed on Earth over a time interval $T' = T(1+z)$. Since less
        photons are observed on Earth per unit time, this decreases the flux by a
        factor of $1+z$.

        On the other hand, a photon emitted with period T has period $T' =
        T(1+z)$ when it arrives on Earth. By the Planck-Einstein relation, the
        photon was emitted with energy

        $$E = \frac{h}{T},$$

        and arrives on Earth with energy

        $$E' = \frac{h}{T'} = \frac{h}{T(1+z)} = \frac{E}{1+z}.$$

        This decreases the flux by an additional factor of $1+z$.

   (b) A circle of radius d on the sphere has a *smaller* perimeter than $2\pi d$. By
        analogy, a spherical surface of radius d on the hypersphere has a *smaller*
        area than $4\pi d^2$. Thus for a hyperspherical Universe the flux at the same
        distance d is *larger* than in the Euclidean case. In other words, a star at a
        certain distance looks *brighter* in a hyperspherical Universe.
        (In hyperbolic space a spherical surface of radius d has a *larger* area than
        $4\pi d^2$, and so the flux at the same distance d is *smaller* than in the
        Euclidean case; thus a star at a certain distance looks *dimmer* in a
        hyperbolic Universe).

9. If the Universe is hyperspherical then the first Friedmann equation is

$$H^2 = \frac{8\pi}{3}(\rho + \rho_\Lambda) - \frac{1}{R^2}.$$

Dividing this equation by

$$H^2 = \frac{8\pi}{3}\rho_c$$

we obtain

$$\frac{\rho + \rho_\Lambda}{\rho_c} - 1 = \frac{1}{H^2 R^2},$$

and so

$$\frac{1}{H^2 R^2} < 0.01 \Leftrightarrow R > \frac{1}{0.1H}.$$

In geometrized units,

$$\frac{1}{H} \simeq \frac{300,000}{71} \text{ megaparsecs} \simeq 4,200 \text{ megaparsecs,}$$

yielding

$$R \gtrsim 42,000 \text{ megaparsecs.}$$

If this were indeed the radius of the Universe, the antipodal point of Earth would be (by analogy with the sphere) at a distance

$$d = \pi \times 42,000 \simeq 132,000 \text{ megaparsecs,}$$

that is, about 10 times the distance to the boundary of the observable Universe, moving away with velocity

$$v = Hd = \frac{H\pi}{0.1H} \simeq 31$$

times the speed of light.

10. If $\dfrac{\Delta R}{\Delta t}$ were constant then the age of the Universe would be

$$\frac{R}{\Delta R / \Delta t} = \frac{1}{H} \simeq 3.26 \times \frac{300,000}{71} \text{ million years} \simeq 14 \text{ billion years}$$

(since if we use years as units then 1 parsec $= 3.26$ years). This value is indeed very close to the exact value (computed taking into account the variation of $\frac{\Delta R}{\Delta t}$ in time).

11. From the first Friedmann equation (with $\Lambda = 0$) we see that for the Euclidean and hyperbolic Universes the expansion would proceed forever, but would not accelerate: the expansion rate would approach zero in the Euclidean case or one in the hyperbolic case. The hyperspherical Universe, however, would attain a maximum radius (equal to $2E$) and collapse again to a zero radius hypersphere (*Big Crunch*).

12. Since the metric of the Einstein Universe is

$$\Delta \tau^2 = \Delta t^2 - R^2 \left(\Delta \theta^2 + \cos^2 \theta \Delta \varphi^2 \right),$$

we see that to maximize $\Delta \tau^2$ the particle must minimize

$$\Delta s^2 = R^2 \left(\Delta \theta^2 + \cos^2 \theta \Delta \varphi^2 \right),$$

that is, must move along a geodesic of the sphere.

Consequently, light rays must also travel along geodesics of the sphere. Therefore any galaxy in the Einstein Universe sees multiple images of itself, each image corresponding to light rays which circumnavigated the Universe a certain number of times. Note that these images show the galaxy at different times, separated by time intervals of $2\pi R$ (where R is the radius of the Universe). Moreover, these light rays arrive from all directions in the sky, and so the corresponding images are magnified. Indeed, the apparent size of any object moving away from the galaxy decreases only until it is at a distance $\frac{\pi}{2}R$ (i.e. until it reaches the "equator"), starting to *increase* from that point on, reaching the maximum as it passes the antipodal point (at a distance πR).

13. (a) The cosmological horizon is due to the existence of the Big Bang: light has only had about 14 billion years to propagate. If the radius of the Universe were constant, this would mean that the cosmological horizon would be 14 billion light-years (that is 4,300 megaparsecs) away; the expansion of the Universe increases this number (because the Universe was smaller in the past).

     (b) The horizon corresponds to points whose light was emitted at the Big Bang, when the radius of the Universe was zero. From the redshift formula in the FLRW models we see that the redshift for these points is infinite.

     (c) Light would take $3.26 \times 14 \simeq 46$ billion years to travel the distance to the horizon.

     (d) This time interval is $\frac{46}{14} \simeq 3.3$ times the age of the Universe. There is no contradiction here, because light *did not* travel this distance to reach us from the horizon, since the Universe was smaller in the past.

14. (a) The value of H is determined by the first Friedmann equation, where the term $\frac{k}{R^2}$ can be ignored (since $\rho + \rho_\Lambda \simeq \rho_c$). As R increases, ρ decreases but ρ_Λ remains constant. Since $\frac{\rho_\Lambda}{\rho_c} \simeq 73\%$, we see that H^2 will always be larger than 73% of its current value, and so H will always be larger than

$$H_\infty = \sqrt{0.73} \times 71 \simeq 61 \text{ km/s/Mpc}.$$

     (b) Let R_0 be the radius of the Universe at a given time t_0. Then for $t > t_0$ we will have

$$\frac{\Delta R}{\Delta t} = RH > R_0 H_\infty,$$

and so R will double in a time interval smaller than

$$\frac{R_0}{R_0 H_\infty} = \frac{1}{H_\infty} \simeq 3.26 \times \frac{300,000}{61} \text{ million years} \simeq 16 \text{ billion years}.$$

     (c) Light being emitted now by a galaxy more than 32 billion light-years away will take at least 16 billion years to get half-way. But meanwhile the radius of the Universe will have doubled, and so this light will find itself again more than 32 billion light-years away from Earth.

     (d) We start by noticing that 32 billion light-years are $\dfrac{32,000}{3.26}$ $\simeq 10,000$ megaparsecs. A galaxy at this distance (as any other galaxy) can only see objects closer than 14,000 megaparsecs, and the light from these objects will never reach the Earth. So objects more than

14,000 + 10,000 = 24,000 megaparsecs away will never be seen from Earth.

15. The sky is dark at night because there exists a cosmological horizon: we cannot see the stars beyond a certain distance. If this were not the case (and the stars were eternal) then we would see infinite stars and the sky would be unbearably bright.

Quantitatively, assuming a static Euclidean Universe containing in average n stars per unit volume, we would have approximately

$$n \times 4\pi r^2 \times \Delta r$$

stars on any spherical shell of radius r and thickness $\Delta r \ll r$. Assuming an average star luminosity L, the total light flux measured on Earth for the spherical shell would be

$$F = n \times 4\pi r^2 \times \Delta r \times \frac{L}{4\pi r^2} = nL\Delta r$$

(independent of r). Thus the total flux would be infinite if there were infinitely many contributing spherical shells.

Even if the Universe were hyperspherical (thus containing finitely many stars), the situation would not change, as in this case one would see infinite images of each star. Consequently the sky is not dark at night in the Einstein Universe.

16. (a) The temperature of the cosmic background radiation is proportional to the energy of its photons. By the Planck-Einstein relation, this energy is inversely proportional to the period, and hence inversely proportional to the radius of the Universe. Therefore the temperature of the cosmic background radiation was infinite at the Big Bang.

   (b) Since the temperature of the cosmic background radiation is inversely proportional to its period, it has suffered a redshift of

$$1 + z = \frac{2940}{2.7} \simeq 1089 \Leftrightarrow z \simeq 1088$$

since it ceased to interact with matter.

17. If the *Enterprise* were in our Universe then it would have gone back in time to the epoch when the radius of the Universe was half the current radius, since the temperature of the cosmic background radiation in the unknown region is twice its current value. In this epoch the vacuum energy density was the same, but the matter density was $2^3 = 8$ times bigger. By the first Friedmann equation (where the term $\frac{k}{R^2}$ can be ignored, since $\rho + \rho_\Lambda \simeq \rho_c$) the square of the Hubble constant would then be

$$0.73 + 8 \times 0.27 = 2.9$$

times bigger, that is, the Hubble constant should be

$$\sqrt{2.9} \times 71 \simeq 121 \text{ km/s/Mpc}.$$

Since the observed value is very different, the *Enterprise* cannot be in our Universe.

Chapter 8
Mathematics and Physics

8.1 Mathematics for General Relativity

What is the mathematics needed to understand general relativity?

First it is necessary to know *infinitesimal calculus*. This basic chapter of mathematics, mostly started by Leibniz[1] (Fig. 8.1) and Newton in the seventeenth century, is used in almost all areas of applied mathematics, from economy to particle physics. In infinitesimal calculus one studies the fundamental concepts of *limit*, *derivative* and *integral*. For instance, when we said that the instantaneous velocity u is the ratio

$$u = \frac{\Delta x}{\Delta t}$$

for very small Δt, we should have said, to be rigorous, that u is the *limit* of this ratio as Δt tends to zero, and should have written

$$u = \lim_{\Delta t \to 0} \frac{\Delta x}{\Delta t}.$$

This limit is actually an example of a *derivative* (instantaneous rate of change). We say that u is the derivative of x in order to t, and write

$$u = \frac{dx}{dt}.$$

On the other hand, when we said that the length l of a curve could be computed approximating the curve by a broken line and adding the length of each segment,

$$l = \Delta s_1 + \Delta s_2 + \cdots + \Delta s_N,$$

---

[1] Gottfried von Leibniz (1646–1716), German mathematician, philosopher, diplomat and lawyer.

J. Natário, *General Relativity Without Calculus*, Undergraduate Lecture Notes in Physics, DOI: 10.1007/978-3-642-21452-3_8,
© Springer-Verlag Berlin Heidelberg 2011

Fig. 8.1 Gottfried von
Leibniz

Fig. 8.1 Gottfried von Leibniz

we should have said, to be rigorous, that l is the *limit* of this sum as the length of the larger segment tends to zero (and so the number of segments tends to infinity). This limit is an example of an *integral*, and is written

$$l = \int ds$$

(the symbol $\int$ is an old form of the letter "s", meaning "sum"). The concepts of derivative and integral are (in a precise sense) inverse of each other, according to the *fundamental theorem of calculus* (discovered by Newton and Leibniz).

As we saw, the motions of particles in general relativity (as well as in the Newtonian theory) are obtained by solving *differential equations*, which are actually equations relating unknown functions to their derivatives. Many physical phenomena (sound, light, heat, waves) are modelled by differential equations (the Einstein equation itself is a complicated set of ten differential equations).

Besides these basic concepts, the rigorous formulation of general relativity requires knowledge of *differential geometry*, the area of mathematics that studies curved spaces.

8.2 Modern Physics

As we saw, it becomes necessary to replace Newtonian physics by special relativity when the speeds involved are comparable to the speed of light, and by general relativity when the gravitational fields involved are strong enough to produce speeds of that magnitude. This does not mean that Newtonian physics is wrong: only that it has a limited range of applicability.

In the same way, it was discovered, also in the beginning of the twentieth century, that it becomes necessary to replace Newtonian physics by *quantum mechanics* when studying very small objects. If in addition speeds are comparable to the speed of light, quantum mechanics must be combined with special relativity; the resulting theory is called *quantum field theory*, and underpins the *standard model*, which describes all known particles and forces except gravity. To include gravity into this framework it would be necessary to combine quantum mechanics with general relativity, thus obtaining a theory of *quantum gravity*; unfortunately, no one has been able to achieve this so far. Currently, the most promising candidate to do so appears to be the so-called *superstring theory*.

Astronomical Data

- Speed of light: 300,000 km/s.
- Radius of the Earth: 6,400 km.
- Radius of the Sun: 2.3 light-seconds.
- Radius of the Moon's orbit: 1.3 light-seconds.
- Radius of the Earth's orbit: 8.3 light-minutes.
- Earth's geometrized mass: 4.5 mm.
- Sun's geometrized mass: 1.5 km.
- Hubble constant: 71 km/s/Mpc.
- Radius of the observable Universe: 14,000 Mpc.
- Age of the Universe: 14 billion years.

J. Natário, *General Relativity Without Calculus*, Undergraduate Lecture
Notes in Physics, DOI: 10.1007/978-3-642-21452-3,
© Springer-Verlag Berlin Heidelberg 2011

Bibliography

Elementary

1. H. Bondi, *Relativity and Common Sense* (Dover publications, Dover, 1986)
2. R. Geroch, *General Relativity from A to B* (Chicago University Press, Chicago, 1981)
3. E. Harrison, *Cosmology: The Science of the Universe* (Cambridge University Press, Cambridge, 2000)
4. L. Sartori, *Understanding Relativity: A Simplified Approach to Einstein's Theories* (University of California Press, California, 1996)
5. B. Schutz, *Gravity from the Ground Up: An Introductory Guide to Gravity and General Relativity* (Cambridge University Press, Cambridge, 2003)
6. R. Stannard, *Relativity: A Very Short Introduction* (Oxford University Press, Oxford, 2008)
7. D. Styer, *Relativity for the Questioning Mind* (Johns Hopkins University Press, Baltimore, 2011)
8. T. Takeuchi, *An Illustrated Guide to Relativity* (Cambridge University Press, Cambridge, 2010)
9. E.F. Taylor and J.A. Wheeler, *Spacetime Physics* (Freeman, New York, 1992)
10. E.F. Taylor and J.A. Wheeler, *Exploring Black Holes: Introduction to General Relativity* (Addison Wesley, San Francisco, 2000)
11. R. Wald, *Space, Time and Gravity: Theory of the Big Bang and Black Holes* (Chicago University Press, Chicago, 1992)

Advanced

12. S. Carrol, *Spacetime and Geometry: An Introduction to General Relativity* (Addison Wesley, San Francisco, 2003)
13. J.B. Hartle, *Gravity: An Introduction to Einstein's General Relativity* (Addison Wesley, San Francisco, 2003)
14. C.W. Misner, K.S. Thorne and J.A. Wheeler, *Gravitation* (Freeman, New York, 1973)
15. B. Schutz, *A First Course in General Relativity* (Cambridge University Press, Cambridge, 2009)
16. R. Wald, *General Relativity* (Chicago University Press, Chicago, 1984)

J. Natário, *General Relativity Without Calculus*, Undergraduate Lecture
Notes in Physics, DOI: 10.1007/978-3-642-21452-3,
© Springer-Verlag Berlin Heidelberg 2011

Webpages

17. A. Hamilton, Inside black holes, http://jila.colorado.edu/ ∼ ajsh/insidebh/index.html
18. E. Wright, Cosmology tutorial, http://www.astro.ucla.edu/ ∼ wright/cosmolog.html

Index

J. Natário, *General Relativity Without Calculus*, Undergraduate Lecture
Notes in Physics, DOI: 10.1007/978-3-642-21452-3,
© Springer-Verlag Berlin Heidelberg 2011

Lightning Source UK Ltd.
Milton Keynes UK
UKOW02n0311301214

243690UK00001B/30/P